Magnetic Resonance in Food Science
An Exciting Future

Edited by

J.-P. Renou
*Research Unit Qualité des Produits Animaux, INRA Theix,
Saint Genes Champanelle, France*

P. S. Belton
School of Chemistry, University of East Anglia, Norwich, UK

G. A. Webb
*c/o Royal Society of Chemistry, Burlington House, Piccadilly, London,
UK*

RSCPublishing

The proceedings of the 10th International Conference on the Applications of Magnetic Resonance in Food Science held in Clermont Ferrand, France, September 13-15 2010.

Special Publication No. 332

ISBN: 978-1-84973-233-8

A catalogue record for this book is available from the British Library

Published by The Royal Society of Chemistry,
Thomas Graham House, Science Park, Milton Road,
Cambridge CB4 0WF, UK

Registered Charity Number 207890

For further information see our web site at www.rsc.org

Printed and bound in Great Britain by CPI Antony Rowe, Chippenham and Eastbourne

Preface

The Clermont-Ferrand-Theix Institut National de la Recherche Agronomique (INRA) was proud to organize the 10th *International Conference on the applications of Magnetic Resonance in Food Science* to celebrate its 10th anniversary.

This scientific event was held from 13 to 15 September 2010 in Clermont-Ferrand. The conference attracted 90 participants from 14 countries from all over the world. The conference included 7 invited lectures, 19 oral presentations and 27 oral poster presentations. Moreover, before the scientific sessions, two postgraduate sessions were given in parallel every morning.

The conference was divided in 6 sessions covering i) Data processing, ii) New developments/food system, iii) New developments/NMR, iv) Nutrition, v) Metabolomic and vi) Imaging. The book follows the form of the conference.

This year's meeting corresponded to its 10th anniversary. The first international conference was held in 1992 at the University of Surrey in Guilford on Peter Belton's and Graham Webb's initiative. It was a great pleasure to listen to Peter's lecture about the past development of magnetic resonance in food and think about where it might go in the future. During the last 20 years, a lot of developments were performed and the next 20 years are also very exciting.

This meeting presentations were focused on the new developments in NMR techniques: hardware as well software with metabolomic and imaging without the new applications of NMR tools in food of course and now in nutrition.

These developments keep pushing further the knowledge limits.

I would like to express my sincere thanks to my co-editors: Peter Belton and Graham Webb for their confidence. I was very glad to organise this conference with the scientific and local committees. This conference was meant to facilitate the communications between participants, giving the opportunity to everybody to get in contact with scientists in the field of magnetic resonance in food science.

The editors hope this 10th anniversary meeting could open new horizons and increase collaborations between scientists involved in application of NMR in food science.

The editors thank all the participants whose work is documented by this book. Additional thanks are given to the production staff of the Royal Society of Chemistry for much helpful advice during the production of this volume.

Dr. Jean-Pierre Renou

Contents

New Developments in NMR

Nutrition

Metabolomics

Imaging

MAGNETIC RESONANCE IN FOOD SCIENCE — TWENTY YEARS FORWARD AND TWENTY YEARS BACK

Peter S Belton

School of Chemistry, University of East Anglia, Norwich, NR4 7TJ, UK

1. INTRODUCTION

Any account of the recent past of magnetic resonance and any attempt to look forward must take account of the deeper history of the subject and consider its essential features. A possible starting place is the beginning of the 19th century in England. At that time the Romantic Movement in the arts was beginning to flower and there was increased interest in the phenomena of electricity and magnetism. The writings of Emmanuel Kant were also beginning to be appreciated in England and among his interpreters was the English poet and man of letters Samuel Taylor Coleridge[1]. Coleridge was an influence on a number of scientists at the time and of particular relevance to this paper is his influence on Michael Faraday. The particular idea was Coleridge's interpretation[1] of Kant's physics to the effect that "all forces are the same". This matched well with the interaction between electric and magnetic forces that Faraday was exploring. The discoveries of Faraday led to the work of James Clerk Maxwell who unified the electric and magnetic fields and in so doing created the idea of electromagnetic radiation: a notion which is critical to the understanding of magnetic resonance. However this in itself is insufficient; as of equal importance in understanding the phenomenon is the idea of quantisation introduced by Planck.
The beginning of the 19th century was a ferment of scientific progress. Nothing that was done at that time can be said to be irrelevant to the development of magnetic resonance. But two events are of particular importance: the discovery of the electron by JJ Thompson and the discovery of the nucleus of Ernest Rutherford. These ideas led to the modern concept of the atom, but one more step was need before all the parts were in place. This was the concept of spin quantum numbers as introduced Wolfgang Pauli. Necessarily this idea contained the implication of nuclear and electronic magnetism. It only needed the development of suitable electronics for the demonstration of the phenomenon of electron spin resonance by Bleaney and Zavoisky and nuclear magnetic resonance by Bloch and Purcell.

2. THE BASIS OF MAGNETIC RESONANCE

As indicted above the phenomenon of magnetic resonance arise as consequence of the spins of electrons and nuclei. The spin generates a magnetic dipole which can align parallel or anti parallel to an imposed magnetic field. The parallel orientation being the lower energy one. The energy difference between the two levels and hence and the energy of radiation need to observe resonance is given by:

$$E = h\nu = \Gamma B_0 \qquad\qquad 1$$

Where E is the energy of the photon needed for the transition, h is Planck's constant Γ is the appropriate constant for the electron or nucleus and B_0 is the applied magnetic field strength. Typically Γ for an electron is of the order of 10^3 times greater than that for a proton. The frequency of resonance varies from typical ranges of 100 to 1000 MegaHertz for proton magnetic resonance and from gigaHertz to hundreds of gigaHertz for electron resonance. This is a very low frequency range compared to the 10^{-13} to 10^{-14} Hertz range of infrared spectroscopy, for example, and it has number of consequences. The relative populations in the upper (N) and lower (N_0) levels, at temperature T, are governed by the Boltzmann equation:

$$N = N_0 \exp(-E/kT) \qquad\qquad 2$$

Where k is the Boltzmann constant.

Table 1 shows the population ratio between the levels for a number of different frequencies.

Frequency	N/N_0
100 MegaHertz	9.9998E-01
1 GigaHertz	9.9984E-01
500 GigaHertz	9.2086E-01
10^{14} Hertz	6.8977E-08

Table 1 *The ratio of populations of the upper and lower energy levels for absorptions at a range of frequencies.*

The table shows that in the radio frequency range even at 500 GigaHertz (currently a typical value for spin echo ESR) the ratio of spins in the upper and lower levels is close to 1. Since the observable signal is proportional to the difference in populations of these levels it demonstrates that magnetic resonance is a very weak effect compared to other spectroscopies such as infrared (10^{14} Hertz) where the difference is very large.

Most textbooks do not discuss the significance of the low energy transition any further but in fact the population difference is the least important aspect of low energy spectroscopy. Low energy quanta are very rare at normal temperatures as is illustrated in Figures 1 and 2. The plots show the radiation intensity at 293K calculated from Planck's radiation equation. As can be seen from Figure 1 the maximum of the radiation is in the infrared region. At low frequencies the intensity is very low, as shown in Figure 2, where the relative intensity is normalised to the maximum emission. This very low natural density of quanta at the frequencies of interest ensures that background radiation is very low and therefore noise generated from the sample itself and the outside world is very small.

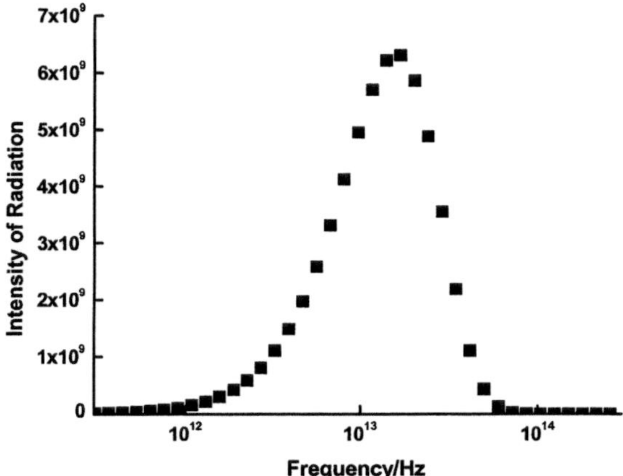

Figure 1 *A plot of thermal radiation intensity versus frequency at 293 K*

In addition to this weak background, the low frequency of the radiation makes the probabilities of spontaneous transitions low and therefore leads to long relaxation times and narrow resonance lines.

Even this, however, is not the most important aspect of the low energy of the transition: low energy quanta make the generation of very large numbers of quanta possible using relatively low energies. A 100 Watt transmitter operating at 300MHz would generate about 10^{17} photons per frequency unit[2]

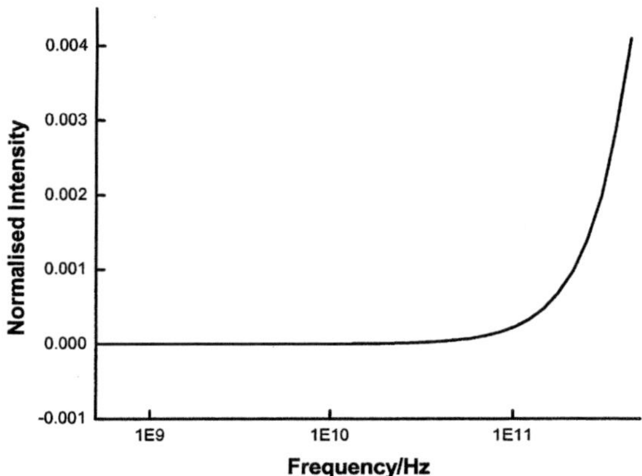

Figure 2 *A detail of Figure 1 showing the very low intensity of radiation at 293K in the radio frequency region*

A form of the uncertainty principle can be written as[3]:

$$\Delta n \Delta \theta \approx 1 \qquad\qquad 3$$

Where Δn is the uncertainty in the number of photons and $\Delta \theta$ is the uncertainty in the phase. Since the number of photons is very large the phase of the system is controlled by the radiation field. Magnetic resonance is therefore coherence spectroscopy. This means that the phase and amplitude of the signal can be controlled by the experimenter and the number of possible experiments is limited only by the ingenuity of the operator. This is in contrast to most optical spectroscopy where the only possible experiment at normal intensities of irradiation is to observe the incoherent absorption of photons.

3. HISTORICAL DEVELOPMENTS

The 1970's to the 1990's saw the general use of pulsed methods in NMR and the consequent exploitation of phase encoding to enable techniques such as 2 dimensional NMR and NMR imaging. It also saw developments in instrumentation with the development of high resolution solid state NMR and larger magnets. The next two decades brought major developments in science which have been reflected in magnetic resonance. These were the massive increase in bioscience, the advent of nanotechnology and the digital revolution. Each of these has impacted in a different way on the field. The response to bioscience has been to the application of magnetic resonance to biological problems. NMR in particular, but also ESR has contributed immensely to our understanding of protein structure and function. The role of NMR in this context has been recognised by the award of a Nobel Prize to Kurt Würtrich. Hyphenated techniques combining NMR and mass spectrometry[4] and applications of proton NMR to metabonomics[5] have reinvented NMR as an analytical technique.

On the whole the development of new magnets has been incremental rather than spectacular: super conducting magnets have increased in field strength and have become more efficient. More innovative developments have been in the application of novel magnets in low field systems. Single sided NMR using the NMR "Mouse"[6] and NMR for online and at line processing using translational relativity through static RF and magnetic fields[7] are interesting and potentially very valuable developments.

The dramatic increase in the availability and capability of digital technology has cause major changes both in the instrumentation and the data analysis. On the instrumental side digital pulse shaping, filtering and oversampling have made advances in experimentation and the quality of data possible. The quality of fast digitisers and other changes have made spin echo ESR a routine, if expensive, technique.

Data analysis may be classified into three different processes:

Extraction: in which data may be manipulated to improve resolution or signal to noise ratios. There may be some selection of subsets of data either on *a priori* grounds or by *a posteri* analysis by such methods as principle components. However the data is treated the overall point of the process is to arrive at a set on numbers that are thought to describe the system. A very comprehensive account is given in the book by Douglas Rutledge[8]

Correlation: in which the relationship between data derived parameters is determined by examination of the relationship of data points or data sets under the influence of same variable parameters or parameters. An example of this is the Laplace transform method[9]

Assignment: In which the data sets are classified as belonging to a particular set. A common example of this classification process is the assignment of sample to an authentic or inauthentic set[10].

Nanotechnology has developed at a very rapid pace. The direct impact of the technology on food has so far been limited but may well become more important in the future. ESR and NMR have been used extensively in the characterisation of systems containing nanostructures but a more intriguing development is the development of nano-scale spectroscopy[11, 12]. Figure 3 shows the set up for atomic force microscope based magnetic resonance detector[11].

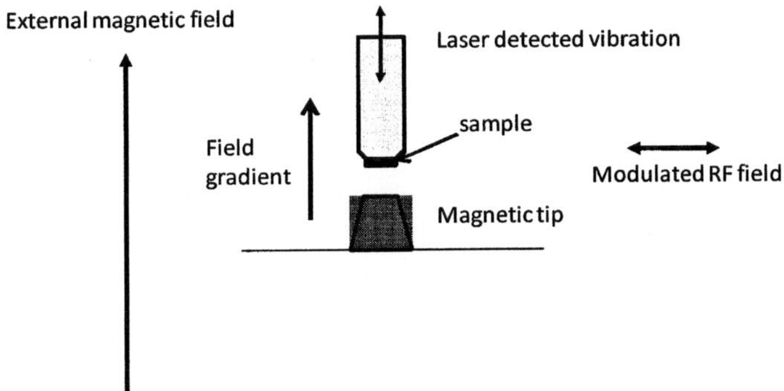

Figure 3 *The configuration for a nano-scale magnetic resonance detector*

The whole apparatus is placed inside a magnet at about the correct field for the radio frequency source to be at the Larmor frequency. The magnetic tip induces a field gradient which means that only part of the sample is on resonance. The sample itself is attached to the cantilever tip of an atomic force microscope which can detect atomic scale displacements. The principle of detection is by directly detecting the movement induced by the creation of nuclear magnetisation. A periodic inversion of the magnetisation is induced by scanning the radio frequency in and out of resonance. Finally, if required, a map of proton density can be produced by the mechanical movement of the tip in the XY plane. An interesting feature of the technique is the since the sample of spins is very small the stochastic variation in magnetisation is greater than the equilibrium Curie magnetisation[12] and this is the magnetisation that is detected. On the scale required it is not currently, possible to build RF coils so a very small strip of wire is used as radio frequency conductor and the RF frequency is ramped at the resonance frequency of the atomic force microscope cantilever. The result is an adiabatic inversion of the magnetisation at the cantilever resonance frequency which amplifies the effect and allows phase sensitive detection. A system has been described[11] which allows imaging at resolution of 10 nm – a 10^8 fold improvement in resolution over conventional imaging methods.

It is interesting to compere this technique with similar developments in infrared spectroscopy[13,14]. Here the use of atomic force microscopy is coupled with a thermal detection system for the absorption of light. This avoids the problem of the diffraction limit for spatial resolution but is limited by the diffusion of heat and the size of the probe. The overall effect of these is that despite infrared being a higher energy spectroscopy the current resolution limit is of the order of 1 to 5 microns.

4. THE FUTURE

It is a truism that predictions of the future are doomed to failure and that the future turns out to be stranger than any of us can imagine. It is also true that the future is a convenient screen on which to project our wishes and prejudices. With these caveats the suggestions made here should perhaps be best seen as a wish list, but one which is at least based in past and current developments.

There are 4 major areas in which one might hope to see change and innovation these are:
- Magnets
- The web
- Bench top machines
- Nano - magnetic resonance

Magnets as normally used in magnetic resonance are large and expensive. This is particularly the case for high resolution NMR. There are two main reasons for this: larger fields result in increased signal to noise ratios and greater dispersion of chemical shift. Higher fields also can simplify second order spectra to first order. This is of considerable value in, for example, protein NMR. However in the increasingly important field of proton NMR for analytical purposes the tendency is to reduce the spectra by "bucketing" in which small regions are co-added to form a histogram. Thus resolution is already deliberately reduced. In the case of low dispersion or poor shimming information about chemical shifts and individual resonances is not lost. It is simply convoluted with a point spread function. Prior knowledge from a single high field experiment may then be used to reconstruct the poorly dispersed spectrum if required. However this may not be necessary, as has been amply demonstrated, by the widespread application of near infrared methods in which spectral resolution of individual absorbances is very rare. There may therefore be a strong case for the use of lower specification and lower field NMR for analytical purposes.

The advances in delocalised computing now available via the web will increase and the question therefore arises: to what extent will local computing for spectroscopy become redundant? Cloud computing offers almost limitless computing power and safe storage. There is no reason why spectrometers may not be run remotely and data handling all be offline.

Magnetic resonance is still an expensive form of spectroscopy, albeit a powerful one. bench top machines have improved considerably over the past decades but their cost is comparable to a top of the range infrared spectrometer, whereas their performance is limited compared to top of the range magnetic resonance instruments. It is unlikely that it will be possible to match such performance on bench top machines but one may hope for continued improvement and for some innovative thinking in magnet design to enable more flexible applications.

The last two decades have seen steady improvement in magnetic resonance rather than the quantum leaps of the previous two decades. Nano-magnetic resonance seems to offer the possibility of such a new innovation. The detection method is base on the direct detection of polarisation rather than the current standard methods of phase coherence. The exploitation of stochastic fluctuations is also novel. There remain problems and the subject is still in its infancy but the future looks very exciting.

References

1. A. I. Miller, *Insights of Genius*, MIT Press, Cambridge , Mass, 1996.
2. P. Belton, *Annual Reports on NMR Spectroscopy*, 1995, **31**, 1-18.
3. A. Abragam, *The Principles of Nuclear Magnetism*, OUP, Oxford, 1978.
4. M. Spraul, U. Braumann, M. Godejohann and M. Hofmann, in *Magnetic Resonance in Food Science - a View to the Future*, eds. G. A. Webb, P. S. Belton, A. M. Gil and I. Delgadillo, 2001, vol. 262, pp. 54-66.
5. J. K. Nicholson, P.J.D. Foxall, E. Holmes, G. H. Nield and J. C. Lindon, in *Magnetic Resonance in Food Science*, eds. P. S. Belton, I. Delgadillo, A. M. Gill and G. A. Webb, Royal Society of Chemistry, Cambridge, 1994, pp. 177-190.
6. G. Guthausen and A. Kamlowski, in *Magnetic Resonance in Food Science: Challenges in a Changing World*, eds. M. Gudjonsdottir, P. Belton and G. Webb, 2009, pp. 46-56.
7. B. Hills, K. Wright, N. Marigheto and D. Hibberd, in *Magnetic Resonance in Food Science: From Molecules to Man*, eds. I. A. Farhat, P. S. Belton and G. A. Webb, 2007, pp. 157-166.
8. D. N. Rutledge, ed., *Signal Treatment and Signal Analysis in NMR*, Elsevier, Amsterdam, 1996.
9. S. Godefroy, L. K. Creamer, P. J. Watkinson and P. T. Callaghan, in *Magnetic Resonance in Food Science: Latest Developments*, eds. P. S. Belton, A. M. Gil, G. A. Webb and D. Rutledge, 2003, pp. 85-92.
10. E. F. Boffo, M. M. C. Ferreira and A. G. Ferreira, in *Magnetic Resonance in Food Science: Challenges in a Changing World*, eds. M. Gudjonsdottir, P. Belton and G.A. Webb, 2009, pp. 143-150.
11. C. L. Degen, M. Poggio, H. J. Mamin, C. T. Rettner and D. Rugar, *Proc. Nat. Acad. Sci.*, 2009, **106**, 1313-1317.
12. S. Kuehn, S. A. Hickman and J. A. Marohn, *J. Chem. Phys.*, 2008, **128**, 052208.
13. X. Dai, J. G. Moffat, A. G. Mayes, M. Reading, D. Q. M. Craig, P. S. Belton and D. B. Grandy, *Anal. Chem.*, 2009, **81**, 6612-6619.
14. J. G. Moffat, A. G. Mayes, P. S. Belton, D. Q. M. Craig and M. Reading, *Anal.Chem.*, 2010, **82**, 91-97.

Data Processing

ADVANCED PROCESSING IN NMR : FOURIER & LAPLACE TRANSFORMS, MODERN SOFTWARE ENVIRONMENT & DATA MANAGEMENT

M.A. Coutouly[1] and M.A. Delsuc[1,2]

[1] NMRTEC, Boulevard Sébastien Brandt, 67400 Illkirch-Graffenstaden France
[2] Institut de Génétique et de Biologie Moléculaire et Cellulaire, INSERM, U596; CNRS, UMR7104; Université de Strasbourg, 1 rue Laurent Fries, 67404 Illkirch-Graffenstaden France

1 INTRODUCTION

In 1965, James W. Cooley and John W. Tukey designed a divide and conquer algorithm allowing a considerable speed-up of the computation of the Digital Fourier Transform[1]. This improvement, named the Fast Fourier transform (or FFT), allowed at that time a 8192 long array to be computed in 8 seconds on a IBM 7094 computer.

A year later, Richard R. Ernst and W.A. Anderson[2-3] proposed the use of the Fourier transform to analyze NMR Free Induction Decays and to produce spectra with a sensitivity incomparable to the one that could be afford by Continuous Wave spectroscopy. The conjunction of computer advances and of understanding of the physical process have marked the opening of a new era for NMR spectroscopy. Algorithmic improvements like the FFT have a strong impact of what is accessible to experimental science and what is not. Without this fortunate conjunction, FT-NMR would have much less successful, and certainly 2D NMR would still be in its infancy.

2 FOURIER AND LAPLACE TRANSFORMS

The Fourier transform possesses many distinctive properties. It is a mathematical operation that keeps the number of data-points unchanged, and thus can be computed in-place. It also has a inverse which is nearly equivalent to the direct operation, and both the direct and the inverse FT can adopt the same fast algorithm.

In a more general approach, processing of an NMR data-set consists in taking numeric data from the spectrometer, and applying to it a series of processing phases in order to deliver a final result which may takes many different aspects : a graphic, a spectrum, a frequency list, a concentration.

Behind this general approach, two different kinds of processing techniques can be recognized depending on what is being modeled : the system under study or the way the measurement is performed[5].

The most common approach consists in a «strong» modeling of the system under study, where a mathematical model is used to describe what is assumed to be known on the sample. This model being parameterized by usually a few parameters, these few parameters are conveniently extracted from the data by some kind of fitting. As the number of fitted parameters is much smaller than the number of acquired points, this is often referred to as «data reduction».

Another usual approach is to model the acquisition process itself rather than the studied system. FT-NMR is a typical example of this approach. Assuming an impulse response, a truncation measurement, and some phase errors, an image of the spectrum is produced after the Fourier transform, with eventually some zero-filling step or phase correction. The mathematical model of the system in this case is minimum or «weak», and mostly associated with the acquisition process (causality, finite spectral-width, finite power, ...) with eventually general assumptions such as the positivity of the spectrum. This approach is called the «transformation» approach, as mathematical transforms are used to produce the final spectrum.

Many such transforms are used in NMR analysis, one can cite Fourier, Hilbert or Hartley in spectral analysis, Haar or wavelet transforms in time-frequency analysis, Hadamard transform in spectral coding and detection, Laplace transform in relaxation and diffusion analysis, Radon transform in MRI and nD NMR, etc.. Most of these transforms are defined as integral transforms over a set of functions, usually designed as mutually orthogonal (the kernel functions). There are generally linear, which is a requisite for a transformation which should produce some kind of «image» from an experimental input. Indeed the sum of the transforms of two independent experiments should be more-or-less equivalent to the transform of the added experiments.

The modeling of the acquisition process is then coded in the way the transform is applied e.g. apodisation and zero-filling in Fourier transform; choice of projection angles in Radon transform, sampled values in Laplace transform.

They also share the characteristic of being computer intensive, and requiring a lot of memory storage, as many of the cited transform do not have the chance of having a «Fast» version as the Fourier transform does. Even worse, many of these transforms do not possess a directly computable inverse, in consequence, it can be quite easy to compute the experimental answer of a model spectrum, but the estimate of the optimum spectrum from the experimental data-set might be very challenging. For example, the Laplace transform which related the relaxation distribution to the shape of the decay of a T_2 echo signal is straightforward to compute, whereas the analysis of the same experimental decay in terms of T_2 distribution is considered as one of the most difficult computational task. This may seem strange to people acquainted only with the Fourier transform, however the same difficulty is observed in the analysis of partially or non regularly sampled FIDs for instance[7].

In all these cases, the image/spectrum generated by the transform contains more independent data points than were initially acquired. This has to be done without adding any «information» to the data, and cannot be performed without some additional *a-priori* assumptions on the final result, like positivity, smoothness or parsimony. These assumptions can be either implicitly embedded in the algorithm, or explicitly handled by an optimization procedure. In the latter case, this is done by creating a target function of the current image/spectrum which mix an estimate of the distance from the experimental data (usually with a chi-square) and an evaluation of the departure from the *a-priori* assumption : total curvature for a maximum smoothness, entropy for parsimony, etc... These independent quantities are then related with a Lagrange multiplier which has also to be optimized. This general approach is called the inverse reconstruction, as one does not have explicitly to compute the spectrum from the data, but only to compute the data that would have been produced by a given spectrum.

One is faced with a formidable computer task, consisting in finding the optimum Lagrange multiplier(s) while minimizing the distance to the experimental data. Each step usually require several direct transform of the whole dataset. With the computers available today, the general case is usually intractable. However at least two very successful cases are already widely used today. The first is to apply this approach on the inverse Laplace transform of small data-sets. The pioneer program CONTIN[8] first introduced this possibility in 1973, and was further extended in 1998 using Maximum Entropy[9]. This allows to process relaxation or diffusion data-sets in a general manner, without any assumption on the content of the sample. This inverse approach can also be successfully implemented over the Fourier transform thanks to this «fast» version. While first proposed for the processing of astronomical data, this approach found its use in NMR on the analysis of non uniformly sampled NMR dataset. The inverse approach here is a real plus, as the lack of certain data-points hinders the possibility to realize a direct analysis.

3 MOORE'S LAW

The same year 1965 the Fast Fourier transform was introduced, Gordon E. Moore, Intel Co-founder, stated that the complexity of electronic component is doubling every year, and could continue to grow at the same rate for at least a decade. Coined as the Moore's law, this trend, revisited to a doubling of performance every 2 years, has since then been verified over more than four decades and shows no sign of speeding down.

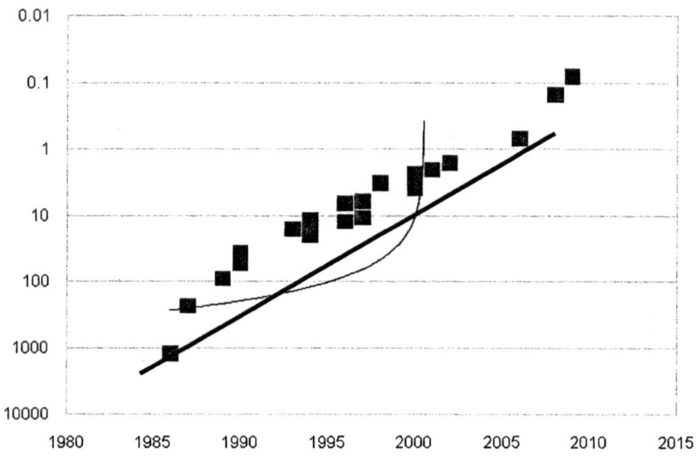

Figure 1 *an example of the Moore's law : processing of a 512 x 2k NMR data set with the Gifa program[6], exponential Broadening and FT in F2; phasing, 5 points Spline base-line correction in F2; cosine apodization, zero-filling and FT in F1; phasing in F1. The first point is a μVaxII running VMS 5.2 in 1986, with a processing time of 1200 sec, the last point is a 3.0 GHz Intel Xeon running MacOs X in 2009, with a processing time of 80msec. The black regression line fits the data on the slope 1000 times every 20 years.*

The fact that computer power doubles every 2 years, or in other words, that computers speed up by a factor 1000 every 20 years should have a profound impact on the scientific method and on the way computers are used.

It has also a strong impact on the way computing methodology can been seen. Indeed a methodology break through, like the FFT, has a long impact on the way scientists and ingeneers work. The FFT used today is still quite similar to the algorithm which was proposed more than 40 years ago. However, the computer on which it runs is several million time faster. This means that a program that runs in 10 minutes today, would have taken 100 years if launched in 1965, and would be only half-done.

With this computer power at hand, and the possibility that the exponential growth does not stop soon, processing becomes a fundamental step in the scientific task. As soon as it requires some mathematical the most common idea is to let a computer do the work. If the problem is a bit tricky and specific, one might consider creating a new software, in other word coding.

4 WHICH COMPUTER LANGUAGE ?

Thinking about writing lines of code, the computer scientist has to insure its software quality. One might think about testing, deployment, documentation and other things that are mainly consumer orientated focus. Of course it is part of the work but, one shouldn't forget the importance of the development quality which enables long term development and software evolution. Team management, revision system and bug tracking are essential tools for any software you want to use at a professional level.

The main pitfall you must avoid is to write a code you are the only one to understand, at the moment you write it. If someone else has to lean on it in order to pursue your work or if you want just to make changes in your code a few months after you wrote it, a lot of time can be wasted just to understand your code. Even though a program is like a book, a form to express your own ideas, if you didn't comment it for YOU while writing it, you might have trouble rereading it. A program always has to be readable and commented correctly so that anyone else knowing the programming language you use, would localize tricks and bugs. If this requirement is not filled, even if the program is correct and executes correctly, no evolution will ever be possible and the software won't probably know a long life expectancy. Readability is often linked to simplicity, perfection is reached when there is nothing to remove anymore. It's only at the point when the code is readable, correct and executed that one can concentrate on making the code running faster, when it's possible. Having all these steps in mind, one can fear beginning such a big thing, fortunately some nice tools exist in order to help at each step of development : code search, modern developer tools, code checking, automatic documentation...[11]

Now that we have clarified the good development practice we will have a look at the different method/programming language that one can use. Asking this type of question to computer scientists you will get some sharp opinions. It can be argued on the pros and cons but one can emphasis some objective parameters to compare them. Assuming that feasibility is not discriminant for any of them, comparing things like time for programming, source text productivity, run time, memory consumption and program length can give a really good picture on the present possibilities.

A battery of evaluations has been lead by Lutz Prechelt in the year 2000[12] on the implementation of an algorithm converting telephone numbers into word strings. He went through the comparison of languages like Perl, Python, Rexx and Tcl that are forming the *script group* (languages that are interpreted and that don't require variable declarations). The other group is *non-script group* which gather more conventional languages like C, C++, Java that are compiled and require typed variable declaration.

The author asked for volunteers to code in a programing language they could choose among the ones that were cited above. At the end of the submission process, the number of usuable program were 80. Those programs have been created by 74 different programmers (mostly students and non professionals), this study is working on the results any average programmer can achieve.

All programs have been executed on the same computer, while their run time show pretty similar value, the memory consumption shows contrasted differences.

Java programs are clearly the least memory-efficient and C/C++ seem to be slightly better than script family. One the other hand, development times show a striking trend where script languages tend to lead to faster development than non)script language. Both groups confirm that the number of line one can write per working hour is rather concern, thus giving the advantage to script languages which have a more concise lay-out. As we have seen when concentrating on development quality, the key point is the way the programmer writes the code. Having a look at the development time, results show that one can develop significantly faster with script languages than with conventional languages.

The study of Prechelt allows to conclude that scripting languages (Perl, Python Rexx and Tcl) are more productive than conventional languages. In terms of run time and memory consumption, they are often better than Java and just as good as C or C++.

Many others studies have been lead in order to achieve the same objective comparison, for instance on http://shootout.alioth.debian.org website, one can find studies on run time of different type of program. In this study, the authors take into account the maximum speed that can be achieved. Typical information one can have is reproduced Figure 2, with C, C++, Java, Python and Perl. The results here are quite contrasted with the ones obtained by Prechelt, as the maximum obtainable speed is concidered, rather than the speed typically obtained with a rapid development.

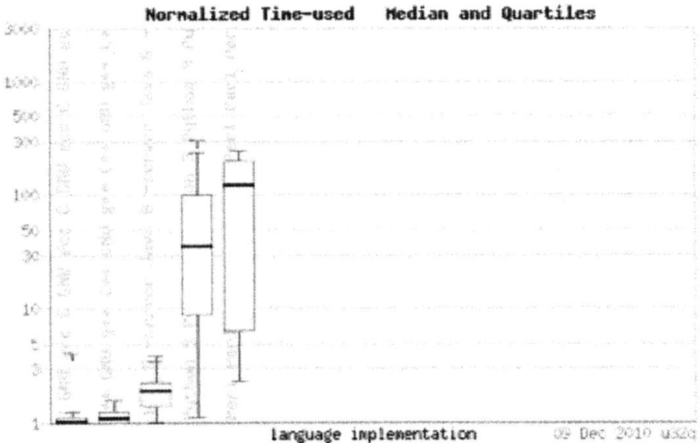

Figure 2 Chart taken from http://shootout.alioth.debian.org website comparing run time of programs in different languages. Each column represents a language, each box represents middle 50% of run time measured. The black line shows the median time measured.

The choice of the programming language is also influenced by the environment in which the programmer works. Having a look at long term trends for the top 10 programming languages in terms of popularity (Figure 3), One striking detail is the apparition of a language in 2009 (Objective-C) that knows a rapid growth as being the language used in some smartphones.

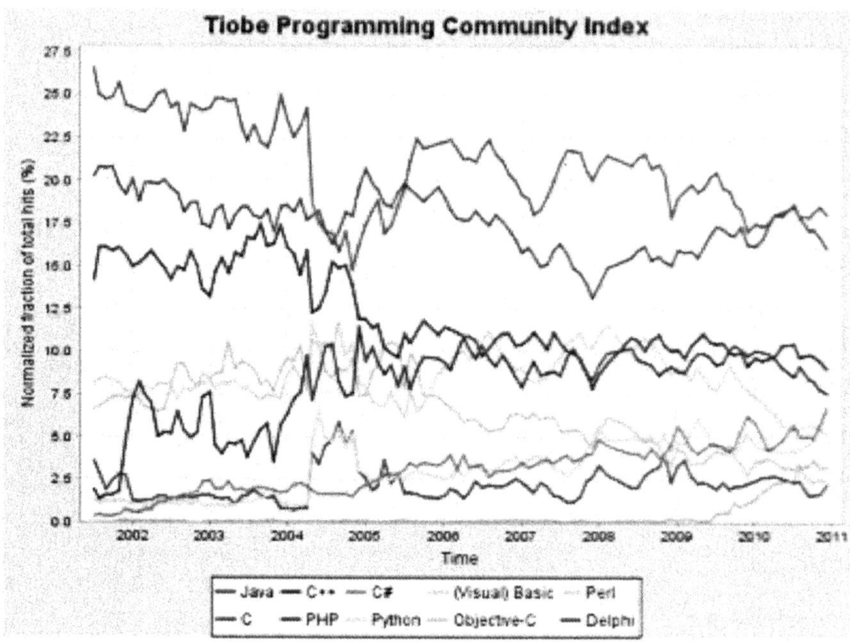

Figure 3 *Long term representation (8 years) of popularity for the top 10 programming language. Python, C#, and Objective C undergo a fast rising of their popularity this year. Chart captured from http://www.tiobe.com/ website.*

Having a more general look, we see that Java seems to be declining, C & C++ seems to be mostly stable whereas Python seems to grow slowly. Interestingly enough, Python and Objective-C are both rising fast enough to win to the programming language of 2010 award. Python is in a very different ecosystem than Objective-C, and we can then assume its popularity is rising for some more profound reasons than the keen interest for some new technology.

5 CONCLUSION

The choice of a programming language can be lead by lots of different factors. Depending on the type of project, the number of time the program will be used, the type of person who will use it, one might want to choose carefully the language having a look at informations such as run-time, memory consumption, source text productivity whereas other will just use the most convenient for them. It also depends on the number of programmers who will have to work together on the code. If only one programmer knows a programming language and is an expert in that language, one has to evaluate how difficult it will be for the other member of the team to learn the given language. Isn't it better to choose the language that everybody knows partially? How easy/difficult will it be teaching the language to the latest student? In general, before coding you should always ask yourself carefully which programming language you want to use. Even the smallest project can, one day require adjustment or expansion and it is always a waste of time having to begin everything from scratch again just because you didn't take enough time for reflection.

Finally, we would like to stress how much a program may be an integral part of the scientific development. As such, we would like to advocate that it is really considered as such. This means to consider programming as seriously as sample preparation or spectrometer choice. It also means that the programs used for a given scientific work are fully part of this work and should be considered in the refereing process. A program code is yet another way of expressing one's scientific ideas and as such should be evaluated and distributed along with the same significance than the other expressions of the scientist. Useless to say that no program code which has an important role in a published work should ever be sealed, but should be made fully available and testable along with rest of the work.

References

1 J.W. Cooley, and J.W. Tukey, *Math. Comput.*, 1965, **19**, 297.
2 R. Ernst, *Adv. Magn. Reson.*, 1966, **2**, 1.
3 R. Ernst, W. Anderson, *Rev. Sci. Instrum.*, 1966, **37**, 93.
4 G.E. Moore, *Electronics Magazine.* 1965, 4.
5 D Tramesel, V Catherinot., MA Delsuc, *J Magn Reson*, 2007, **188**, 56
6 JL Pons, TE Malliavin, MA Delsuc, *J Biomol NMR* 1996, **8**, 445
7 MR Gryk, J Vyas, MW Maciejewski, *Progress in Nuclear Magnetic Resonance Spectroscopy*, 2010, **56**, 329
8 S.W. Provencher, *Compt. Phys. Commun.* 1982, **27**, 213
9 MA. Delsuc and T.E Malliavin, *Anal Chem*, 1998, **70**, 2146
10 D. Rovnyak and JC. Hoch and AS. Stern and G. Wagner, *J Biomol NMR*, 2004, **30**, 1
11 N. Chauvat, Euroscipy 2010, Paris, July 8-11 2010
12 L. Prechelt, *IEEE Computer* 2000, **33**, 23

USING PARAFAC CORE-CONSISTENCY TO ESTIMATE THE NUMBER OF COMPONENTS IN LF-NMR DATA - APPLICATION TO IN-SITU STUDIES OF MECHANICALLY INDUCED GEL SYNERESIS IN CHEESE PRODUCTION

C.L. Hansen, F. van den Berg and S.B. Engelsen

Quality & Technology, Department of Food Science, Faculty of Life Sciences, University of Copenhagen, Rolighedsvej 30, 1958 Frederiksberg C, Denmark

1 INTRODUCTION

A major challenge with low field NMR data analysis is estimation of the right number of exponential components. In this work we will demonstrate how PARAFAC core-consistency, using the so-called DoubleSlicing method *(1)*, can help to unambiguously determine the number of exponentials in the signal. The new approach will be demonstrated in a study where we investigate in-situ rennet induced milk gel formation and in-situ mechanically induced gel syneresis using low field nuclear magnetic resonance (LF-NMR) *(2)*.

In cheese manufacture the milk gel formation and syneresis processes are of major importance for the water content, texture and flavour properties of the final product. Several NMR relaxation studies have been reported that investigate milk gel formation *(3-6)* and syneresis *(4, 6, 7)*. The effect of milk gel syneresis on water proton relaxation has so far been studied on undisturbed gels that only exhibit spontaneous endogenous syneresis (i.e. syneresis caused by pressure being built up during network formation within the gel). This is however not representative for industrial cheese manufacturing where mechanical cutting of the gel into dices is an essential process step. Two inconsistencies exist in the interpretation of the water proton relaxation during milk gel formation and syneresis: (A) the development of the transverse relaxation time constant, T_2, and its corresponding proton population size during gel formation and (B) the number of proton populations (i.e. exponential terms) necessary to model water proton relaxation during cheese gel formation. Two studies have found no change in the relaxation time constant T_2 during milk gel formation *(4, 6)*, while other researchers *(3)* found small changes in the T_2 relaxation time during the gel formation. The latter study found (without providing explicit proof) that three populations of water protons were required to describe the relaxation during milk gel formation, while the former studies found that one proton population was adequate to describe the proton relaxation. There is however a general agreement that the onset of milk gel syneresis is associated with appearance of an additional water population with slower relaxation, which are the protons in the whey water *(2-4, 6)*.
One of the reasons for the discrepancy in number of proton components is related to the data analytical methods applied for studying the NMR relaxation data. The major challenge when analysing relaxation decay curves of LF-NMR experiments using multi-

exponential curve fitting is to decide the appropriate number of exponential terms that describe actual water populations present in the sample. A new exponential curve fitting method to assist in the determination of water populations called DoubleSlicing was introduced in 2003 by Micklander et al. *(8)*. Andrade et al. *(1)* further refined the method using the PowerSlicing scheme of Engelsen and Bro *(9)*. The DoubleSlicing technique utilizes the fact that different parts (slices) of a given multi-exponential decay curve consist of the same underlying exponential terms, but in a different quantities (concentrations). Because all underlying exponential terms are present in all slices of the DoubleSliced data cube, tensor models such as direct tri-linear decomposition (DTLD) or parallel factor analysis (PARAFAC *(10)*) can extract the individual exponential terms. These tensor methods have some unique possibilities for validating the solution (e.g. finding the appropriate number of water proton populations) and are surprisingly faster than conventional curve fitting methods.

The primary objective of this study was to investigate the effect of milk gel formation and in-situ mechanically induced gel syneresis on NMR proton relaxation. The secondary objective was to demonstrate DoubleSlicing as a method for determining the appropriate number of components in a semi-automated way. For these purposes the effect of milk gel formation and syneresis was studied using an experimental design with three factors: pH, temperature and gel firmness at cutting time. Time domain LF-NMR measurements were carried out in parallel with rheological measurements.

2 MATERIALS AND METHODS

2.1 Design of experiment

Rennet induced gel formation of skim milk and the subsequent syneresis process after gel cutting was studied by time domain LF-NMR. As experimental procedure rennet was added to skim milk in a bigger volume which was subsequently split into two fractions: one was immediately transferred into a 17mm diameter NMR tube and continuously analyzed inside a LF-NMR spectrometer; the other fraction was injected into a rheological instrument which continuously measured gel firmness during gel formation. The role of the rheological measurements was to ensure that the gels formed in the NMR tube for repeated experiments had the same, desired firmness at gel cutting time. A knife - consisting of a thin Plexiglass blade tightly matching the tube inner diameter - was used to cut the gel once over the entire length of the NMR tube as soon as the firmness of the gel in the twin sample had reached a pre-defined level (Figure 1).

During the gel formation and syneresis three experimental factors were investigated on two levels (a 2^3 factorial design). A pseudo centre point was added for the factors pH and gel firmness at cutting for the high temperature level (Figure 2). Each of the corner points in the design were run in duplicate, while the pseudo centre point was run in four replicates, resulting in a total of 20 experiments, carried out in random order.

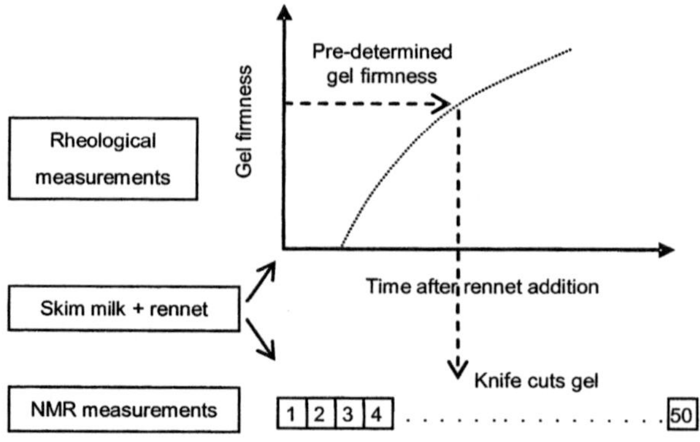

Figure 1   *Timing diagram of the study: Rennet was added to skim milk, which was split into a fraction for rheological analysis and a fraction for LF-NMR analysis. A knife cut the formed gel once over the entire length and inner diameter of the NMR tube, once the gel reached a pre-determined firmness. The LF-NMR spectrometer continued to analyse the gel, during the syneresis, which resulted in fifty snapshot/time-frames for each experimental run.*

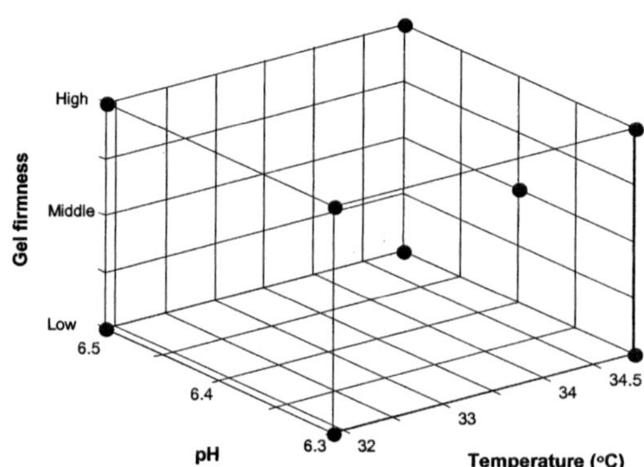

Figure 2   *The randomized 2^3 factorial design used to study gel formation and syneresis. The experimental factors varied were pH, temperature and gel firmness at cutting. The corner points in the design were run in duplicate, while the pseudo centre point (position selected due to experimental limitations) was run in four replicates, resulting in 20 experiments.*

2.2 NMR Spectroscopy

Time domain LF-NMR analysis was carried out on a benchtop 23.2 MHz Maran pulsed ^{1}H NMR spectrometer (Oxford Instrument, United Kingdom) equipped with a 17mm diameter variable temperature probe head. The CPMG (Carr-Purcell-Meiboom-Gill) pulse sequence was used to determine the relaxation behaviour. It was chosen because it minimizes the influence of magnetic field inhomogeneities, diffusion and chemical exchange *(11)*. A total of 8100 data points/echo times were acquired, with a 90–180 pulse spacing (τ) value of 500µs. Only the even numbered data points were used in the data analysis, resulting in 4050 data acquisition points per measurement. Prior to the first measurement the frequency of the instrument was adjusted on a 10mM CuSO4 standard sample. During gel formation four scans were accumulated with a relaxation delay between consecutive scans of 14 seconds. Prior to the four scans each measurement was preceded by two dummy scans leading to a total measurement time of 2 minutes and 12 seconds. Measurements were carried out continuously until a maximum of 100 minutes after cutting.

2.3 NMR data analysis by DoubleSlicing

Time domain LF-NMR data are commonly analyzed using multi-exponential fitting which applies non-linear iterative curve-fitting algorithms to extract and characterize the underlying pure exponentials from random noise in the data (Equation 1):

$$M(t) = \sum_{n=1}^{N} M_{0n} \cdot \exp\left(\frac{-t}{T_{2,n}}\right) + e(t) \tag{1}$$

M(t) is the reduced magnetization at time t, M_{0n} is the concentration or magnitude parameter of the nth exponential, $T_{2,n}$ is the corresponding transverse relaxation time constant and e(t) is the residual error. One of the pitfalls of curve fitting based on hard modelling to a functional form such as Equation 1 is that adding additional exponential components will per definition improve the fit (i.e. reduce the residual) even if only meaningless noise/none systematic variation is being fitted.

Micklander et al. *(1)* introduced an alternative non-iterative and rapid technique for curve resolution called DoubleSlicing. The technique pseudo-upgrades the single relaxation curve to become tri-linear data, by cutting the relaxation curve into slices (Figure 3). By selectively removing parts of the signal curve (slicing) and using the remaining curve along with the original curve, the relaxation curve can be transformed from a one-dimensional signal (a vector) into two-dimensional data (a matrix). By repeating this procedure on the matrix, the data is transformed to three-dimensional data (a cube or tensor). A DoubleSliced relaxation curve can be decomposed using tensor models such as DTLD or PARAFAC into (Equation 2):

$$\underline{X} = ABC^{T} + \underline{E} \tag{2}$$

A, **B** and **C** are matrices and there outer product forms a model/approximation of **X**, while **E** holds the residual variation not explained by the model. If the **X** is decomposed using the right number of components, then matrix **B** will contain the true underlying mono-exponential components *(10)*. From the resolved mono-exponential components it is easy to determine T_2 and M_0 by using Equation 1.

Some major advantages in using tensor models are that they require no initial guesses and have unique possibilities for validation of the solution *(10, 12)*. Methods of solution diagnostics include relative reduction in root mean square error (RMSE), split-half validation and core-consistency. One general challenge of the DoubleSlicing method is to capture the fast relaxing protons, as these are rapidly attenuated in the relaxation curve. Andrade et al. *(1)* therefore further refined the DoubleSlicing method using the

PowerSlicing scheme by Engelsen and Bro *(9)*. The PowerSlicing approach ensures that components with fast relaxing protons are present in sufficiently many of the slices by slicing more frequently at the short echo times, since the fast relaxing component only have a strong presence in the beginning of the curve (first echo times) .

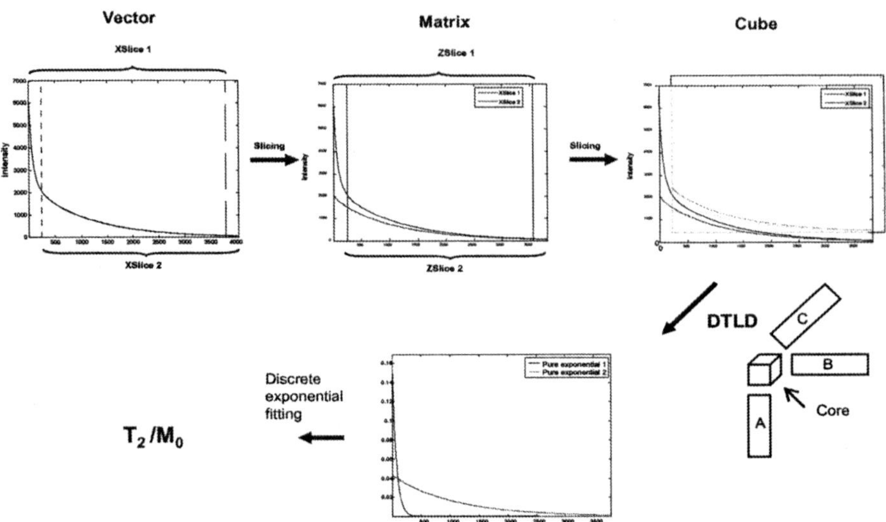

Figure 3 *Overview of the principle behind DoubleSlicing. The NMR signal is one-dimensional data (a vector) and a sum of N exponential decays corresponding to N water populations. The curve is divided in a number (here two for illustration purposes) of largely overlapping segments (X-slices) by removing the same number of first or last echo times. These two segments are placed in one matrix (2-dimentional) and the procedure is repeated to form (Z-slices) which are stacked behind each other to form a data cube (3-dimensional). This cube is decomposed by a DTLD factor model and the correct rank or dimensionality is then equal to N, the number of distinct water populations in the original signal. The profiles are from the DTLD decomposition (so-called loadings) are mono-exponential and can be used to estimate concentrations (M_0) and relaxation time constants (T_2).*

In the present study a step-wise DoubleSlicing algorithm was used as an extension to the Slicing algorithm by Pedersen et al. *(13)*. The DoubleSlicing algorithm slices the single relaxation curve 11 times at 1, 2, 4, 8, 16, 32, 64, 128, 256, 512, 1024 echo time variable (hence eleven X-slices compared to just two shown in Figure 3). The exponential increase in the slicing variable number is the so-called PowerSlicing approach and it ensures that fast relaxing components are present in sufficiently many slices. Subsequently, the eleven slices are PowerSliced once again eleven times, which transform the single relaxation curve into a three-dimensional data cube. The three-dimensional data cube was decomposed by DTLD using one to four components. All the algorithms are available in MatLab code from www.models.life.ku.dk.

2.4 Estimation of the number of exponential components

Many methods have been evaluated for the determination of the appropriate number of exponential terms in Equation 1 or components in the tri-linear decomposition model. Obvious diagnostics criteria are loss in residual, explained variance, etc. However, the task remain difficult using these methods because the fit will always improve by adding more terms and the statistical or numerical evaluation on whether an additional term is justified is far from trivial, often giving conflicting information for different criteria. In this work we developed a semi-automated selection procedure for model complexity based on two-times-two evaluation criterion.

Bro and Kiers *(12)* proved that the core-consistency is a useful validation diagnostic for evaluating the appropriateness of fit of tri-linear models (Figure 4). The core-consistency is used to evaluate the tensor model in Equation 2. It expresses how close or far a model is from the assumed tri-linear structure. It is obtained by comparing the elements in an unconstrained or free core tensor with the element in a constrained (super-diagonal) tensor core (our target core). The core-consistency is expressed as the percentage , where a high percentage indicates that the unconstrained solution is close to or consistent with our desired modeling objective *(12)*. Thus, for a perfect tri-linear model the core-consistency is 100%, the desirable value, whereas e.g. negative percentages indicate a very poor model. Once the maximal appropriate number of components is exceeded, the core-consistency will typically drop dramatically. Core-consistency can thus be used as a diagnostic tool but it should always be used in combination with other diagnostics, because it sometimes leads to over-parameterized models. We therefore combine core-consistency with the loss of fit (RMSE).

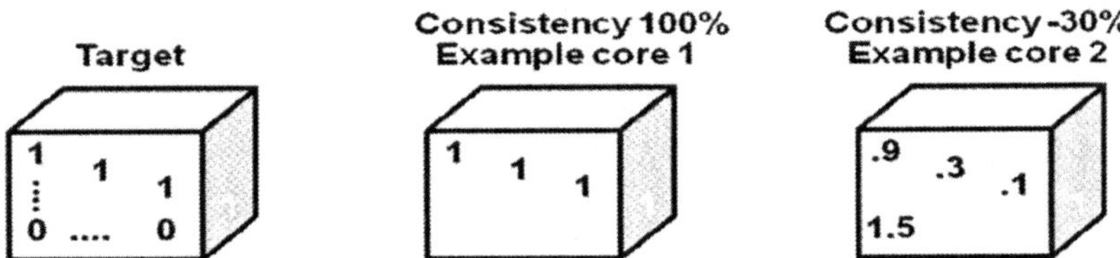

Figure 4   *Tri-linear core-consistency - if the core-consistency is much lower than 100% the tri-linear model is not appropriate i.e. it has the wrong number of components.*

In our model evaluation we construct tri-linear DTLD models with one to four components. Note that e.g. the first component is not the same for these four choices because tensor models are not embedded *(10)*. To justify inclusion of an additional component both criteria (Table 1) should be met. An advantage of the algorithm is that it can automatically find the appearance of additional components during the development of dynamic experiments such as gel formation and syneresis without any prior knowledge on the system. Also note that each LF-NMR relaxation measurement is modelled independently (Figure 3). Hence any observation made over the time axis of the experiment can be based on sovereign measurement points, i.e. single relaxation curves.

Table 1   *Diagnostic methods and their corresponding threshold values used in the algorithm to validate the rank/number of components appropriate to describe each DoubleSliced relaxation curve.*

Diagnostic method	Threshold value for inclusion
Core consistency	> 60%
Loss of root-mean-squared-error[a] (RMSE)	> 10%

[a]Loss in root-mean-squared-error is relative to the model with one component less

3 RESULTS AND DISCUSSION

The raw CPMG data for a representative cheese batch is shown in Figure 5. The data show that the overall relaxation becomes systematically slower with experiment time after cutting of the milk gel. This implies that the proton populations progressively change as a result of the syneresis process, where water (i.e. whey) is expelled from the gel network. By visual inspection of the relaxation curves in Figure 5 it is not possible to observe any changes in the relaxation curves prior to cutting of the gel.

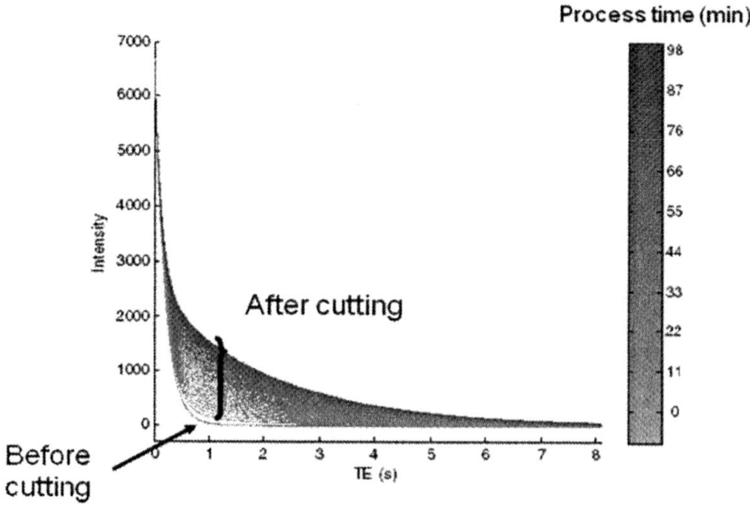

Figure 5   *NMR CPMG relaxation curves during gel formation and syneresis in one gel formation and syneresis experiment.*

The core-consistency and RMSE will be illustrated and evaluated using relaxation curves from one gel formation and syneresis experiment (pH = 6.3, T = 35°C, low firmness at cut). Figure 6 shows how the core-consistency and loss in RMSE change as a function of the number of exponential terms included in the modeling of the relaxation curves. Since core-consistency is consistently 100% when the relaxation curves were fitted to two exponentials, it is clear that at least two populations are present during the entire experimental run. Using three exponentials results in negative core-consistency during the initial phase, gel formation. The change from a core-consistency of 100% to negative values makes it easy to asses that only two components are present during gel formation, prior to cutting. In comparison, the loss in RMSE does not show strikingly different values when including three components. Mono-exponential fitting of the relaxation curves by DTLD before cutting yielded a RMSE of 0.045, while bi-exponential fitting results in a RMSE of 0.038 corresponding to a 15% reduction. Tri-exponential fitting resulted in a RMSE of 0.036 or a 4% reduction from a two component model (i.e. < 10%).

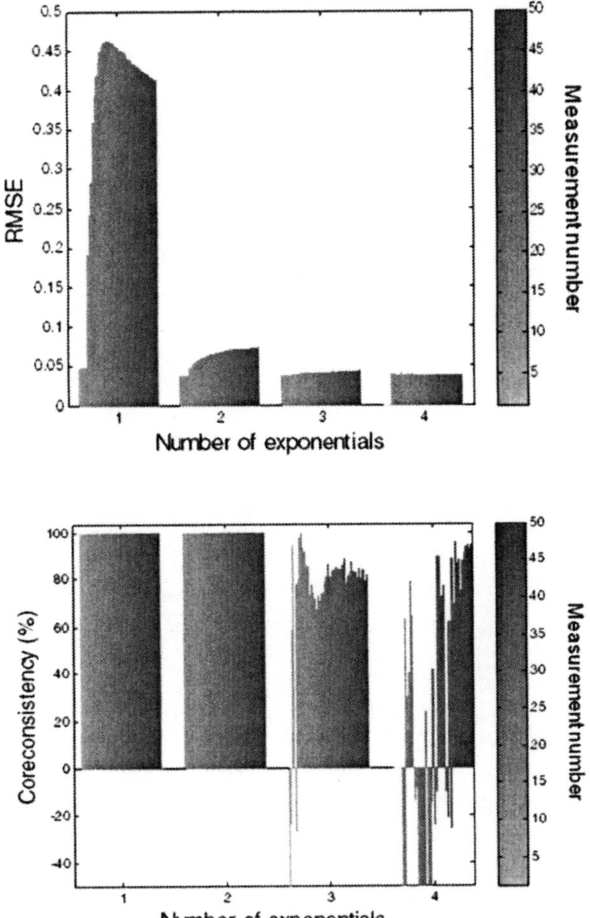

Figure 6   *Core-consistency and loss in RMSE change as a function of batch run-time and the number of exponential terms included in the modelling of a representative cheese batch (consisting of 50 NMR relaxation curves equals one batch of approximately 100 minutes).*

By comparing of the results of DoubleSlicing with traditional discrete exponential fitting of one relaxation curve before cutting (Figure 7) we see that two proton components are in fact present in milk and the gel. This is evident from a plot of residuals after fitting one component (Figure 7B), which show that residuals are not random and equally distributed around zero as LF-NMR measurement noise should be. After fitting two components the residual become random (Figure 7C) showing that no more information is present in the data.

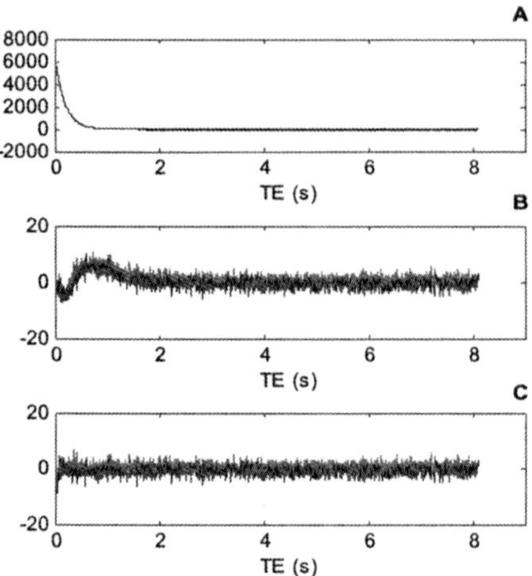

Figure 7   *(A) One CPMG relaxation curve, (B) residual after one component discrete exponential fitting and (C) residual after two components.*

When the appropriate number of proton components has been established the development in T_2 and population sizes can be calculated for each LF-NMR snapshot. The result for two representative batches is shown in Figure 8. The abrupt change at the cutting point and the otherwise smooth curves gives rise to high confidence in both the modelling approach and the automated method of model complexity determination. During gel formation and syneresis two proton populations with the characteristic transverse relaxation times $T_{2,1}$ and $T_{2,2}$ are present within the gel. The size of $T_{2,1}$ and $T_{2,2}$ show that proton population come from water associated with different parts/constituents of the gel. The data analysis clearly indicates that bi-exponential behaviour characterizes the system during gel formation and prior to cutting. Cutting unambiguously introduces a new component $T_{2,3}$ representing the whey.

The relative population sizes determined during modelling quantitatively shows how much water (protons) with different T_2's is present at a given time during the syneresis (Figure 8). The main portion of the (water) protons (relative contribution in total signal of 98.8%, SD = 0.2%) prior to cutting of the gel originate from a population which is characterized by an average $T_{2,1}$ of 180.7ms (SD = 5.1ms). The primary development can be summarized as follows: fast relaxing water ($T_{2,1} \sim$ 180ms) within the gel is mainly converted to slow relaxing water ($T_{2,3} \sim$ 2000 to 2200ms) situated outside the gel. A small fraction of the fast relaxing water protons is seemingly converted into water present within the gel with an intermediate relaxation rate ($T_{2,2} \sim$ 400 to 500ms). The rate by which the water leaves a cut milk gel (i.e. syneresis) can be described as a reaction of first order, meaning that the rate of water expulsion is dependent on the amount of water present at a given time *(14)*. During syneresis the bulk proton population characterized $T_{2,1}$ steadily decrease to a level of $\sim$ 50 – 70% of the water protons after 100 minutes. Assuming, this proton population is primarily associated to casein micelles as previously suggested *(15)*, then the decrease in population size initiated by cutting, suggest that the casein micelles are being steadily dehydrated.

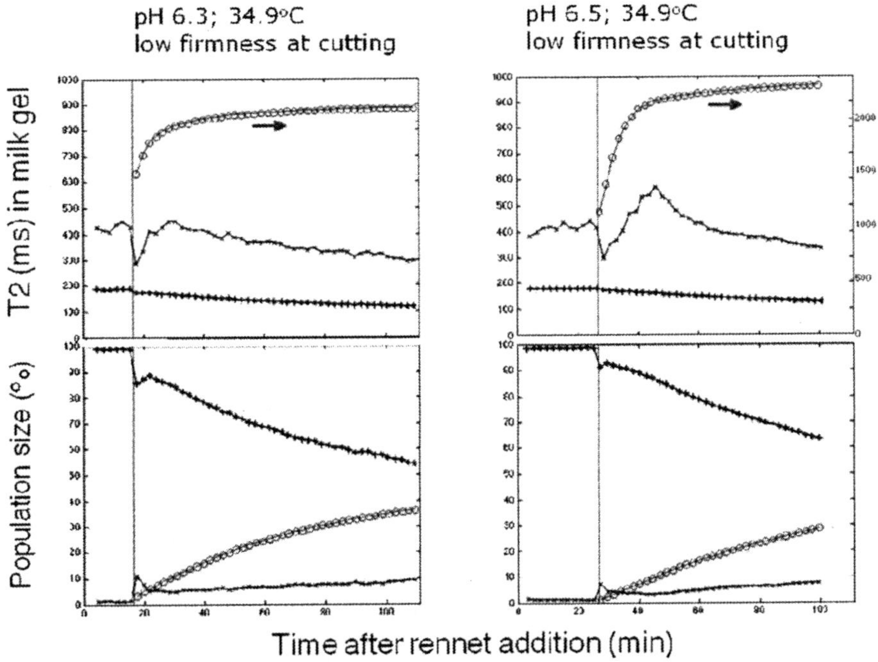

Figure 8 *Development in T_2 (upper row) and relative population size (lower row) during gel formation and syneresis of two different cheese experiments with different pH-values. The vertical line indicates the time when the milk gel was cut.*

4 CONCLUSION

LF-NMR was used to characterize skim milk gel formation and syneresis qualitatively and quantitatively using a new automated DoubleSlicing algorithm. Analysis of relaxation data using DoubleSlicing data proved to be precise in finding the appropriate number of underlying exponential components (i.e. proton populations) in single relaxation curves measured during gel formation and syneresis.

In-situ LF-NMR measurements have proven an excellent tool for studying rennet coagulation and syneresis. During coagulation two proton populations with distinct transverse relaxation times ($T_{2,1}$ = 181ms, $T_{2,2}$ = 465ms) were present in fractions (98.9% and 1.1% respectively). Mechanical cutting of the gel in the NMR tube induced macro-syneresis, which led to the appearance of an additional proton population ($T_{2,3}$ = 1500 to 2200ms) identified as whey. The syneresis rate was found to be significantly dependent on pH in the range from 6.3 to 6.5 and temperature in the range from 32 to 35°C. Gel firmness at cutting did not show any significant effect on syneresis rate *(2)*.

References

1. Andrade, L.; Micklander, E.; Farhat, I.; Bro, R.; Engelsen, S. B. DOUBLESLICING: A non-iterative single profile multi-exponential curve resolution procedure - Application to time-domain NMR transverse relaxation data. *J. Magn. Reson.* **2007,** *189* (2), 286-292.

2. Hansen, C. L.; Rinnan, A.; Engelsen, S. B.; Janhoj, T.; Micklander, E.; Andersen, U.; van den Berg, F. Effect of Gel Firmness at Cutting Time, pH, and Temperature on Rennet Coagulation and Syneresis: An in situ H-1 NMR Relaxation Study. *J. Agric. Food Chem.* **2010,** *58* (1), 513-519.

3. Hinrichs, R.; Bulca, S.; Kulozik, U. Water mobility during renneting and acid coagulation of casein solutions: a differentiated low-resolution nuclear magnetic resonance analysis. *International Journal of Dairy Technology* **2007,** *60* (1), 37-43.

4. Lelièvre, J.; Creamer, L. K. NMR-study of formation and syneresis of renneted milk gels. *Milchwissenshaft-Milk Science International* **1978,** *33* (2), 73-76.

5. Mariette, F.; Maignan, P.; Marchal, P. NMR relaxometry: A sensor for monitoring acidification of milk. *Analusis* **1997,** *25* (1), M24-M27.

6. Tellier, C.; Mariette, F.; Guillement, J. P.; Marchal, P. Evolution of water proton nuclear magnetic relaxation during milk coagulation and syneresis - structural implications. *J. Agric. Food Chem.* **1993,** *41* (12), 2259-2266.

7. Ozilgen, M.; Kauten, R. J. NMR analysis and modelling of shrinkage and whey expulsion in rennet curd. *Process Biochemistry* **1994,** *29* (5), 373-379.

8. Micklander, E.; Thygesen, L. G.; Pedersen, H. T.; van den Berg, F.; Bro, R.; Engelsen, S. B. Multivariate analysis of time domain NMR signals in relation to food quality. In *Magnetic Resonance in Food Science: Latest Developments* , Webb, G. A.; Belton, P. S.; Rutledge, D. N., Eds.; The Royal Society of Chemistry: 2003; pp 239-254.

9. Engelsen, S. B.; Bro, R. PowerSlicing. *J. Magn. Reson.* **2003,** *163* (1), 192-197.

10. Bro, R. PARAFAC. Tutorial and applications. *Chemom. Intell. Lab. Syst.* **1997,** *38,* 149-171.

11. Hills, B. P.; Takacs, S. F.; Belton, P. S. A new interpretation of proton NMR relaxation-time measurements of water in food. *Food Chem.* **1990,** *37* (2), 95-111.

12. Bro, R.; Kiers, H. A. L. A new efficient method for determining the number of components in PARAFAC models. *J. Chemom.* **2003,** *17* (5), 274-286.

13. Pedersen, H. T.; Bro, R.; Engelsen, S. B. Towards rapid and unique curve resolution of low-field NMR relaxation data: Trilinear SLICING versus two-dimensional curve fitting. *J. Magn. Reson.* **2002,** *157* (1), 141-155.

14. Fox, P. F.; McSweeney, P. L. H. *Dairy Chemistry and Biochemistry;* Klüwer Academic/Plenum Publishers: New York, 1998.

15. Le Dean, A.; Mariette, F.; Marin, M. H-1 nuclear magnetic resonance relaxometry study of water state in milk protein mixtures. *J. Agric. Food Chem.* **2004,** *52* (17), 5449-5455.

ISOTOPIC ANALYSIS AND ^{1}H-NMR SPECTROSCOPY FOR TRACEABILITY AND DISCRIMINATION OF ITALIAN WINES

Costanza Aghemo[a*], Andrea Albertino[a] and Roberto Gobetto[a]

[a]*Dipartimento di Chimica I.F.M., Università di Torino, via Giuria 7, 10125 Torino, Italy*
*Author to whom correspondence should be addressed

1 INTRODUCTION

Deuterium Nuclear Magnetic Resonance and Mass Spectrometry applied on carbon-13 have been reported to be reliable techniques for classifying natural products according to their geographical origin[1]. Furthermore the same techniques have been extensively used for detecting sophistications. In literature, samples have been generally classified taking into account wide regions of production.[2] We enlarged the knowledge in this area by focusing our attention to the investigation of the isotopic contents restricted to a more local environment.

SNIF-NMR (Site-Specific Isotopic Fractionation-Nuclear Magnetic Resonance) have been applied on a series of Nebbiolo and Barbera wine samples of different vintages, coming from different areas within Piedmont region (Northern Italy) and the results have been compared with a large set of meteorological parameters recorded on weather stations placed in fields where grapes grew up. Several climatic parameters at local level have been correlated with the isotopic data with the goal of finding the meteorological factors that are linearly dependent to the isotopic content.

On the other hand, the chemical composition of wines has been also investigated using by High-Resolution ^{1}H-NMR spectroscopy, coupled with chemometrics studies, in order to identify peculiarity due to specific production areas, vintages and variety. In literature several papers have been reported on this subject. [3-6]

Therefore, wine samples typical of two different varieties and coming from Piedmont were preliminary discriminate using by High-Resolution ^{1}H-NMR spectroscopy, coupled with PCA, in order to classify wines on the basis of their polyphenolic fingerprint.[5]

In order to investigate phenolic composition, wine samples were collected over different vintages and analysed by using ^{1}H-NMR-PCA approach, after pre-concentration of samples on SPE (Solid Phase Extraction) cartridge.

2 MATERIALS AND METHODS

2.1 Wine Samples for Isotopic analysis

Nineteen samples of wines coming from Piedmont (Northern Italy) and collected over different vintages were analysed. After opening the samples were stored at -20°C and the internal atmosphere was pumped out using nitrogen flow.

2.2 Wine Samples for ^{1}H NMR analysis

Twenty-four independent samples of wines of Piedmont region were investigated: 12 samples typical of Nebbiolo grape variety and 12 samples typical of Barbera variety. After opening, the samples were stored at -20°C and the internal atmosphere was pumped out using nitrogen flow

All solvents used were of analytical reagent grade and were purified according to published procedures. Methanol, ethyl acetate were purchased from Sigma–Aldrich, sodium-3-trimethylsilylpropionate-d4 (TSP-d4), was purchased from Cambridge Isotope Laboratories Inc. and used as received without further purification.

2.3 ^{2}H-NMR Measurements

For isotopic analysis, the sample preparation before isotopic analysis consists of ethanol extraction from wine (stored in the dark at 18°C) and determination of residual water by Karl-Fisher titration, according to the EC Regulation.[7]

NMR spectra on deuterium were registered referring to the certified procedure[7] with a Bruker Avance 600 spectrometer operating at 14 Tesla equipped with a 5 mm broad band probe at 29°C. Samples were prepared by mixing and weighting 0.393 ml of extracted ethanol and 0.157 ml of TMU. Transients were performed without lock and with NOE suppression on proton (inverse-gated decoupling sequence). Spectra were acquired with a 90° pulse (180 μs, 10 dB), spectral width of 1200 Hz (13 ppm) and a 6.8 s acquisition time. Irradiation offset was fixed between OD and CHD resonances whereas the decoupling offset was located in the middle of the frequency interval existing between the CH_3- and CH_2- groups from the proton spectrum measured on the same tube. The number of transients was chosen in such a way to obtain a signal to noise ratio greater than 150 with an exponential factor of 2 Hz applied on FID. Five spectra were recorded on each extracted sample. Therefore, isotopic parameters, $(D/H)_I$, $(D/H)_{II}$ and R, were calculated as previously reported[7] with an exponential factor applied on FID of 0.5 Hz. The parameters calculated are:

$$R = 3\frac{h_{II}}{h_I}$$

where h_{II} e h_I are respectively methylenic and methylic peak heights;

$$(D/H)_I = 1,5866 \cdot T_I \cdot \frac{m_{st}}{m_A} \cdot \frac{(D/H)_{st}}{t} \tag{1}$$

$$(D/H)_{II} = 2,3799 \cdot T_{II} \cdot \frac{m_{st}}{m_A} \cdot \frac{(D/H)_{st}}{t}$$

where T_I is the ratio between methylic and TMU peak heights; T_{II} is the ratio between methylenic and TMU peak heights; m_A and mst are ethanol and TMU weights; t is the title of ethanol; $(D/H)_{st}$ is TMU isotopic ratio supplied by BCR.

2.4 Meteorological data

The meteorological data have been daily recorded by permanent weather stations placed very close to the vineyard. In particular, station in Castiglione Falletto (CN) is sited in the same field in which grapes grew up. The registered parameters were temperature, rainfall and air humidity; by means of these data we calculated mean temperatures, total rainfalls, mean air humidity and several thermal sums concerning the ripening period.

2.5 ^{1}H-NMR measurements

Since phenolic compounds are present in low concentration in wine, a sample pre-treatment is useful for their extraction and concentration before the analysis. Therefore, the polyphenols were extracted from wine samples using reversed-phase (RP) ODS-C18 classic cartridges (BakerbondSpeTM, J.T. Baker, Deventer, The Netherlands), according to the procedure reported by Maraschin *et al.* (2003).[8]
EtOAc-extracted fractions were resuspended in 0.700 ml of acetone-d_6. As polyphenolic compounds have a low solubility in acetone, the solid residue was dissolved in 100 µl of D_2O and the solution was transferred in the little cone. After centrifugation (6000-rpm/5 min), 550 µl of the supernatant were transferred into a 5 mm NMR tube. 0.5 µg of TSP-d_4 were added as internal chemical shift standard.
^{1}H NMR spectra were recorded at 21°C using a Bruker Avance 600 spectrometer operating at 600.13 MHz for ^{1}H and equipped with a 5 mm broad-band probe. To acquire the ^{1}H spectra of wine samples, a pre-saturation sequence (zgcppr) was used to suppress the water signal; the presaturation offset was set on the water signal. Each spectrum was acquired with 1532 scans, a spectral width of 20.0 ppm, an acquisition time of 8 s and a relaxation delay of 1 s. Prior to Fourier Transformation the FID was apodized using a Lorentzian line broadening of 0.5 Hz. Each wine sample was analysed four times and values were averaged. Chemical shifts were referenced to TSP-d_4 at 0.0 ppm.

3 RESULTS AND DISCUSSION

3.1 ^{2}H Results

Meteorological parameters have been correlated with isotopic data. Since ethanol is the main product of sugar fermentation, consequently the direct consequence of photosynthetic cycle, we took into account meteorological data of the short time of the year related to the period of sugar accumulation and grape ripening. This period corresponds to the "veraison", the phenological phase relating to the colour changing of the grapes, and ends with the vintage. Taking into account previous investigations as a starting point of our study and by means of the weather stations located directly or nearby in the grape fields, we were able to carefully investigate if rainfall and mean temperature could afford linear dependence with isotopic abundances and if other weather parameters could be correlated with isotopic data.

It is known that the R parameters identify precursor plant of sugar that forms ethanol in wine after fermentation.[9] For example, ethanol obtained from fermentation of beet or cane sugar, shows R values of 2.7 and 2.2 respectively, whereas not chaptalized wine displays values ranging from 2.40 and 2.65. As supposed, our data (not showed), range within that interval, from 2.477 to 2.636. As concerning the methylic site, samples never showed values lower than 98 ppm, which represents the Italian minimum limit.[10]

In Figure 1 the dependence between deuterium abundance of methylic site (D/H)$_I$ with overall rainfall and mean temperature for each vintage are reported.

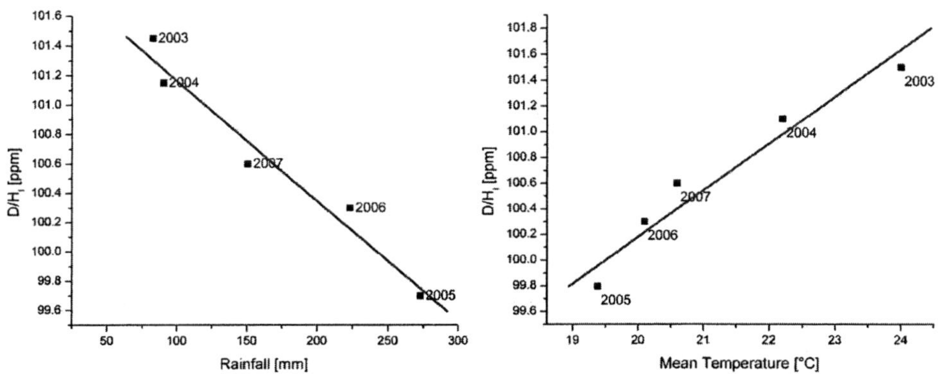

Figure 1   *Correlations between (D/H)$_I$ with rainfall and mean temperature.*
a)   (D/H)I = -0.01P + 102.10; R2 = 0.99; b) (D/H)I = 0.33T + 93.54; R2 = 0.97,
where P is the overall rainfall and T the mean temperature.

Concerning the precipitation, the trend is similar to that reported in previously published papers[11] with the improvement of the reliability of meteorological parameters. Moreover significant linear correlations in inverse proportion, (R^2 close to unit for Nebbiolo wines) have been found. The higher (D/H)$_I$ values correspond to Nebbiolo samples of 2003, the least rainy year taken into account (82.6 mm). In the contrast, samples of 2005, in which the amount of the rainfall was 272.8 mm, show lowest isotopic ratios. It is worth noting that the same slope (0.01) and a very similar intercept (102.10 for Nebbiolo wine and 102.03 for Barbera wine) have been found for different grape varieties.

Taking into account the mean temperature, the trend is in accordance with previous results too[11], because isotopic values are proportionally correlated to the mean temperature of each period. Moreover, we demonstrate that mean temperature and (D/H)$_I$ are in linear direct proportion, with a lower correlation coefficient than rainfall versus (D/H)$_I$ (R^2= 0.97 and R^2= 0.94 for Nebbiolo and Barbera wine respectively).

3.2 ^{1}H Preliminary Results

Principal Component Analysis (PCA) was performed on the NMR data set. The spectral region 5.80-8.40 ppm was segmented into 65 spectral domains having a width of 0.04 ppm (buckets) and each domain was integrated. The integral of the region 5.8-8.4 ppm was set to 1.

PCA was performed on all 65 variables. The PC1/PC2 scores plot explained the 78% of the total variance. The score plot of the principal component (PC1 and PC2) was shown in Figure 2. Excellent separation was obtained for the two grape varieties along a single axis (PC1).

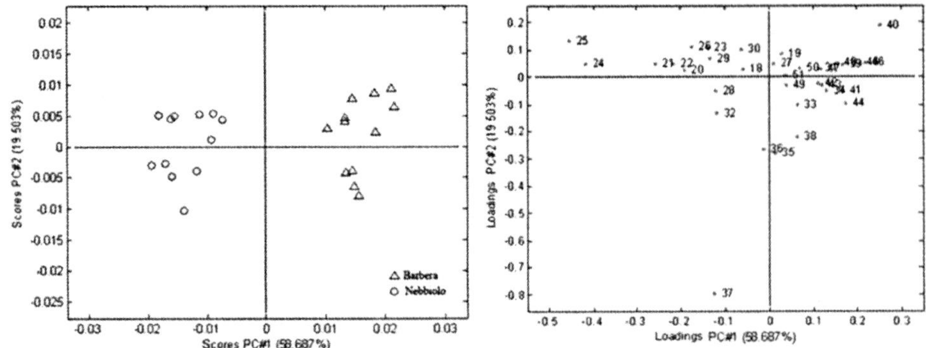

Figure 2   *2D representation of PCA calculated from ^{1}H NMR data:* (○) *Nebbiolo samples,* (△) *Barbera samples. Scores and loadings are reported.*

The associated loading plot allowed identifying the discriminating bucketed data. The buckets responsible of the separation correlated with variable names allowed to the individuation of the spectral regions more discriminating.
In general, the wines of Barbera grape variety were characterized by a positive score on PC1 whereas the wines of the Nebbiolo had a negative score on PC1.

4 CONCLUSIONS

Isotopic parameters such as D/H site specific ratios were investigated and correlated with meteorological data obtained by weather stations placed near vineyards where grapes grew up. Correlations between isotopic data and several climatic parameters at a local level were attempted and a couple of linear correlations were found. Mean temperature and total rainfall were proved to be correlated in linear direct and inverse proportion with the isotopic abundance respectively.

Moreover, wine samples were investigated by using High-Resolution ^{1}H-NMR spectroscopy in order to obtain the classification of Nebbiolo and Barbera wines on the basis of their aromatic fingerprint. Excellent preliminary classification of the variety of the wines was obtained using the NMR-PCA approach.
Our preliminary results suggest the possibility of discriminating wines of different vineyards and sample of the same vineyards coming from distinct areas within the same region. This observation, if confirmed by future studies, will represent a very important tool for a better traceability of the wine origin. Further investigation of the isotopic contents and of polyphenolic fingerprints on several other vineyards are under investigation in our laboratory.

References

1   G.J. Martin, M.L. Martin, F. Mabon, M.J. Michon, *J Agric Food Chem,* 1983, **31**, 311.
2   A. Hermann, *Mitteil Klostern,* 2003, **53**: 132.
3   H.G. Da Silva Neto, J.P.B Da Silva, G.E. Pereira, F. Hallwassa, *Magn. Res. Chem.* 2009, **47**, S127.

4   M. Anastasiadi, A. Zira, P. Magiatis, S.A. Haroutounian, A.L. Skaltsounis, Mirkos, E., *J. Agric. Food Chem.* 2009, **57**, 11067.

5   H.S. G.S. Son, H.J. Hwang, Ahn, W.M. Park, C.H. Lee, Y.S. Hong, *Food Res. Intern.*, 2009, **42**, 1483.

6   G.E Pereira, J.P. Gaudillére, C. Van Leeuwen, G. Hilbert, M. Macourt, C. Deborde, A. Moing, D. Rolin, *J. Int. Sci. Vigne Vin*, 2007, **41**, 103.

7   EC Regulation 2676/90, *Official Journal of European Communities*, 1990, **L 272**, 64.

8   R. P. Maraschin, C. Ianssen, J. L. Arsego, L. S. Capel, A. M. A. Cimadon, C. Zanus, M. S. B. Caro, M. Maraschin, *Magnetic Resonance in Food Science: Latest Developments*, 2003, ed. by Belton, Gil, Webb and Rutledge, The Royal Society of Chemistry, Cambridge, 255.

9   N. Christoph, A. Rossmann, S. Voerkelius, *Mitteil Klostern*, 2003, **53**, 23.

10   Versini G, Monetti A and Reniero F, *Monitoring authenticity and regional origin of wines by natural stable isotope ratios analysis. Wine – Nutritional and therapeutic benefits.* 1997, Watkins, New York, 115.

11   N. Ogrinc, I.J. Kosir, M. Kocjancic, J. Kidrič, *J Agric Food Chem*, 2001, **49**, 1432.

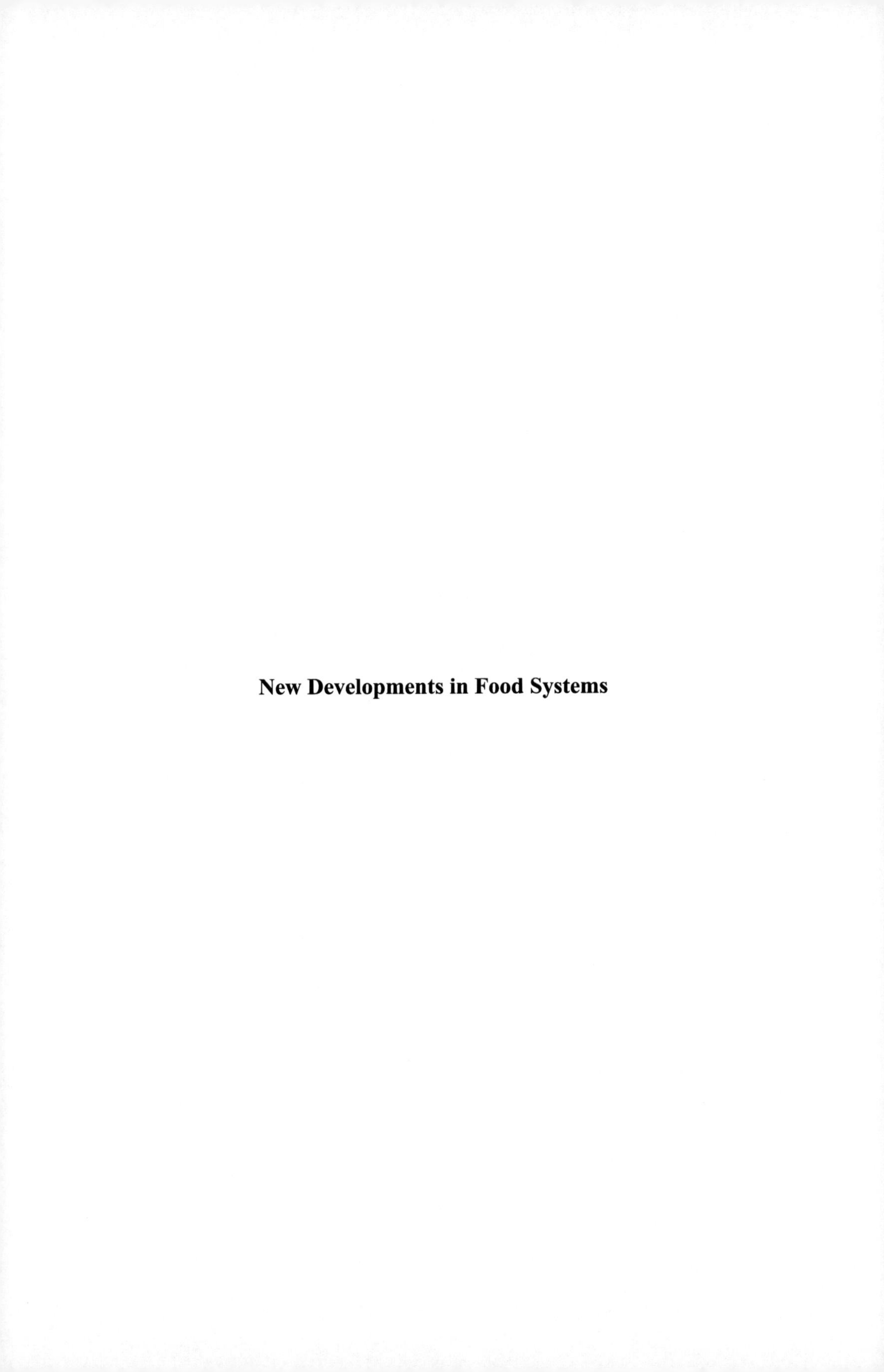

New Developments in Food Systems

PFG-NMR ON DOUBLE EMULSIONS: A DETAILED LOOK INTO MOLECULAR PROCESSES

R. Bernewitz[1], X. Guan[1], G. Guthausen[1], F. Wolf[2] and H. P. Schuchmann[2]

KIT, D-76131 Karlsruhe, Germany
[1]SRG10-2, Institute for Mechanical Engineering and Mechanics
[2]Institute of Process Engineering in Life Sciences, Section I: Food Process Engineering

1 INTRODUCTION

Single emulsions are of high interest in different fields and therefore well investigated. They provide more advantages when an emulsion is part of another emulsion, as double- or multiple emulsion. Multiple emulsions, as novel food systems, reduce the amount of fat in food or encapsulate active substances. They are not only of interest for food, but also for pharmaceutical or cosmetic products.

As in case of single emulsions, multi-modality and droplet size distribution (DSD) of multiple emulsions are important parameters for product quality and physical properties (e.g. viscosity and shelf life). These parameters are also important to gain insight into mechanisms taking place in the second step of the two-step production of double emulsions, which still is not understood in detail. Currently, common measurement techniques for the determination of DSD and its modality of single emulsions are Laser light scattering, Laser diffraction, Ultra sonic attenuation and PFG-NMR. However, acoustic and optical methods do not work properly or not even at all, when aiming for information about structure properties of the inner emulsions. Additionally, results on the outer emulsion are not reliable anymore when using the optical methods, due to multiple diffraction inside the double emulsion.

PFG-NMR is a well established method to determine the droplet size distribution of single emulsions. It is known to work non-destructive with a small workload and a high selectivity. In this context, high selectivity means, that the optical properties of the media do not play a role. Using ^{1}H-NMR for example, only molecules containing hydrogen can be detected. These molecules can be differentiated by their specific chemical shift in high resolution NMR, or by relaxation filters as applied in low field NMR.

Therefore, the PFG-NMR method is in principle suitable to determine structural parameters like DSD in double emulsions. However, the current restrictions have to be exploited, and new models have to be created to analyze the signal decays measured on double emulsions. Molecular processes like diffusion phenomena, relaxation and spectral properties have to be understood in order to obtain a correct interpretation of the data and quantitative meaningful double emulsion's DSD.

2 METHOD AND RESULTS

2.1 Methods

2.1.1 Preparation of W_1/O/W_2 Double Emulsions. The double emulsions were produced in a two-step process: First, the inner emulsion was produced, using a tooth rim dispersing device. The inner emulsion consisted of water W_1 with 0.5% gelatine and vegetable oil with 10% PGPR90. In the second step, the inner emulsion was further emulsified by a colloid mill with water W_2 containing 2% LEO-10 and, recipe depended, 0.6% Xanthan. The typical range of mean oil droplets size is $d_{50,3} = 14$ µm, and of W_1 droplets $d_{50,3} = 6$ µm.
In this work, double emulsions of the type W_1/O/W_2 with a disperse phase ratio of 15%/35%/50% (Figure 1a)) were investigated by PFG-STE NMR experiments (Figure 1b)) and confocal laser scanning microscopy (CLSM).

2.1.2 PFG-STE NMR on W_1/O/W_2 Double Emulsions. The NMR experiments were performed on a Bruker Avance 200 SWB. The parameters of the PFG-STE sequence are summarized in Table 1. The sequence is known to be a suitable method to measure molecular diffusion phenomena. In comparison to a PFG-SE Hahn-Echo experiment, it allows longer diffusion times Δ, due to the "T_2-bypass" after the second gradient pulse so that the diffusion time is restricted by T_1. PFG-NMR measures the displacement of molecules during Δ. In pure substances, the diffusion of molecules is described by random movement, with a final displacement $\Delta r(\Delta)$,[1,2]. In confining geometries, like in an emulsion's droplets, the mean free path differs from the one observed at free self-diffusion. Therefore, molecules exposed to restricted diffusion exhibit a different PFG-NMR signal. Summarizing, molecules in confined geometries show a different diffusion behaviour than molecules in the pure substance.
NMR data of single emulsions are most commonly evaluated with the model of *Murday and Cotts / von Neuman* (MC)[3,4]. This model is valid for single emulsions of water and oil, containing spherical droplets with a mean diameter $d_{50,3} > 1$ µm, assuming the Gaussian phase approximation. The question is, if data of double emulsions are modelled correctly by this model.

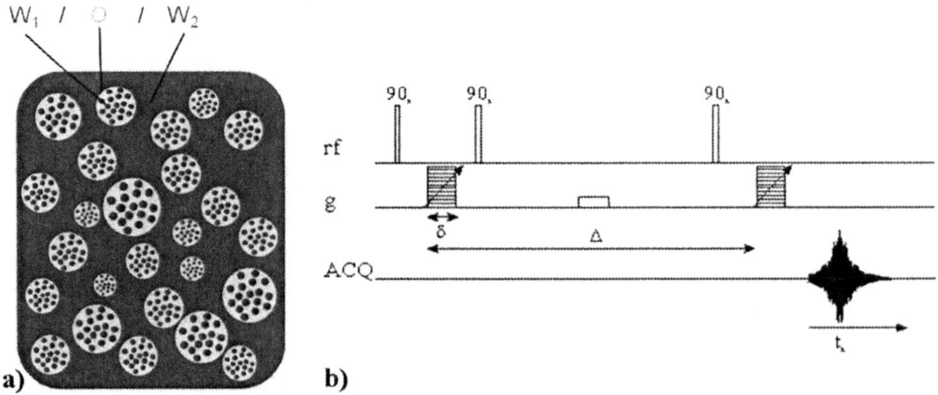

Figure 1 **a)** *Scheme of a W_1/O/W_2 double emulsion* **b)** *PFG-STE NMR sequence.*

Table 1 *Parameters and their values of NMR experiments on a Bruker Avance 200*

Symbol	Parameter	Typical Value(s)
δ	Gradient pulse duration	4 ms
Δ	Diffusion time	50, 100, 150, 200, 300, 500 ms
P_{90}	90° rf-pulse length	10 μs at 0 dB
τ_1	First rf-pulse delay	6.115 ms
D_1	Repetition time	8 s
g	Amplitude of magnetic gradient	3.54 mT/m… 5.99 T/m linear in 32 steps; 3.54 mT/m… 22 mT/m (W_2)
G	Amplitude of spoiler gradient	0.2 T/m
	Duration of spoiler gradient	5 ms
	Rise time for gradient	100 μs
	Time for gradient stabilisation	2 ms
NS	Number of scans	2
DS	Number of dummy scans	2
$\omega_0 = 2\pi\nu_0$	Larmor-Frequency	$2\pi*200.145$ MHz
	Spectral width	20161 Hz
	Number of complex data points	4096

2.1.3 CLSM on $W_1/O/W_2$ Double Emulsions. The double emulsions were additionally investigated by a CLSM in order to compare NMR results to a more direct method. To obtain a contrast between the three phases the oil phase was dyed with Rhodamine-B. CLSM was used because of its 2.5D-capability. This allows a scan of the sample slice by slice.

2.2 Results

2.2.1 NMR Signal Attenuation of Double Emulsions. The spectra of a double emulsion, measured with $\Delta = 0.3$ s, exhibit specific resonances of water (at 4.6 ppm) and oil (Figure 2a)). The investigated double emulsion $W_1/O/W_2$ was prepared as described in *2.1.1* with the speciality that W_2 didn't contain Xanthan. The spectrum amplitudes depend on the gradient amplitude as it is expected for signal from diffusing molecules. In Figure 2b) the logarithmic signal attenuation of the double emulsion is shown together with the logarithmic signal attenuations of W_1 and W_1/O single emulsion, which were retained during the production process.

The pure water phase W_1 only shows a signal near $g = 0$ T/m (Figure 2 b), •). This is due to free self-diffusion of water molecules in W_1. The diffusion coefficient is in accordance with the water self diffusion at the measurement temperature of 22°C.

The NMR data of the single emulsion, shown in figure 2b), was modelled with the MC-model. Assuming a log-normal distribution function for the DSD, it revealed a median $d_{50,3} \approx 3.5$ μm with a width of $\sigma \approx 0.5$ of the corresponding Gaussian distribution. By a Coulter Counter LS230 laser diffraction device, the median of the single emulsion's DSD was determined to $d_{50,3} \approx 2$ μm. Comparing both results, the difference of $d_{50,3}$ is within the mean standard deviation specified on the Coulter Counter.

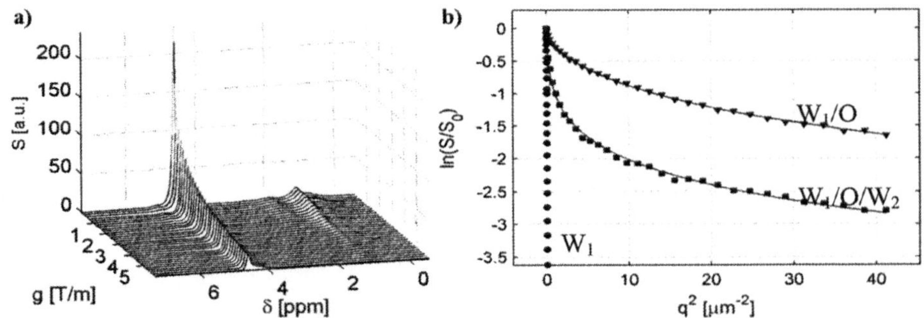

Figure 2 a) *Spectrum of a $W_1/O/W_2$ double emulsion as a function of g.* **b)** *Logarithmic signal attenuation of pure inner water phase W_1, the W_1/O single emulsion after first emulsion step and the double emulsion.*

However, the signal attenuation of the double emulsion differs significantly from the attenuation of the single emulsion. Modelling of the data with the MC-model leads to $d_{50,3} \approx 7$ µm with a width of $\sigma \approx 0.75$. Assuming a complete preservation of the simple emulsion's properties this DSD is physically meaningless. In the analysis, the whole water signal was considered and lead to the wrong DSD.

But there are possible reasons for the failure of the data modelling:

- Also fractions of W_2 following the laws of restricted diffusion could have influenced the signal attenuation.

- Diffusion between the two water phases could occur, which leads to an imprecise separation of the signal of the two water phases. This means, diffusion takes place through the oil phase, which is called molecular exchange in our context. The bounds of the oil droplets function as a thin membrane in this picture. This phenomenon is also described for different inhomogeneous systems[5-10].

2.2.2 Influence of restriction on water diffusion in W_2. In Figure 3, the signal attenuations of O/W single emulsions are shown as a function of the disperse phase concentration. The emulsions were measured via PFG-STE-NMR at 200 MHz with $\Delta = 0.05$ s. The disperse phase ratio was varied from 0% to 50%. 0% means pure continuous phase, consisting of water with 2% Leo10 and 0.2% Xanthan.

The logarithmic signal attenuation of the pure continuous phase is a linear function of q^2, which was expected because of free self-diffusion. The curvature of $\ln(S/S_0)$ increases with growing disperse phase ratio, because of increasing non-spherical restriction of the water molecules between the oil droplets. However, taking the small range of q^2 into account, the influence of this effect is marginal. Regarding the range of q^2 of common PFG-NMR experiments (Figure 2b), the effect can be neglected for W_1 diffusional behaviour.

Concluding, the outer phases in single emulsions as well as the outermost phase in double emulsions have to be considered as following restricted diffusion even at very small disperse phase concentrations.

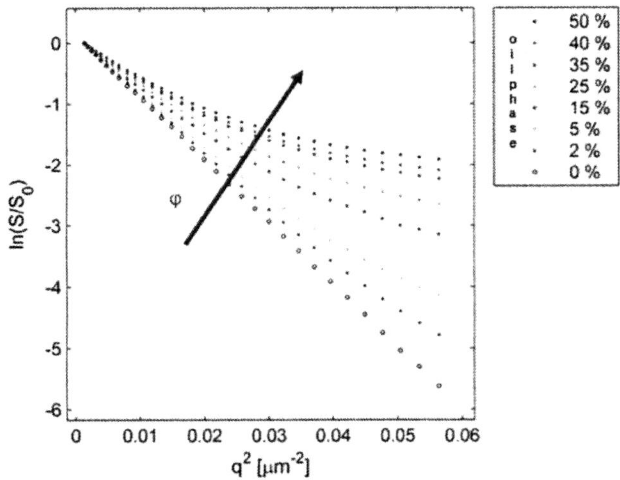

Figure 3 *Signal attenuation of O/W emulsions with different disperse phase ratios φ.*

2.2.3. Diffusion Phenomena between W_1 and W_2 in $W_1/O/W_2$. In several publications, the possibility of molecular exchange in double emulsions is discussed[5-12]. In some publications[8-10], a molecular exchange model was applied, which originally was found to be applicable in case of cellular systems[5-7]. There is, however, some controversy whether such effects are observed or other phenomena are responsible for the experimental results. One argument against the exchange hypothesis is that the solubility of the two substances is low and they tend to separate when mixed.

To find out if and on which time scale diffusion between W_1 and W_2 takes place[11], the idea was to add a small amount of Xanthan to W_2. Xanthan is a common thickener, also used in food industry. It increases the viscosity and consequently decreases the mobility of molecules. In Figure 4a) NMR data obtained from a double emulsion with Xanthan in W_2 is shown. In comparison to a double emulsion without Xanthan (Figure 2a)), a sudden drop occurs after the first gradient step. This difference could be explained by decreased exchange between W_1 and W_2. To verify this suspicion, another experiment was performed, using driven diffusion effects.

Right before the NMR experiment, the $W_1/O/W_2$ double emulsions with Xanthan in W_2 were diluted with additional 50% W_2, containing different amounts of sugar (1%, 5%, 15%). To avoid further large energy input, the additional phase was added by hand while stirring with a glass stick.

As shown in (Figure 4 b)-d)), the signals at $g > 0$ T/m disappear with increasing sugar concentration. By adding sugar, the equilibrium of exchange was influenced, and driven diffusion from W_1 to W_2 took place. The process continues on a short time scale until a new equilibrium between the two phases is established. Regarding this fact, the conclusion is that, despite of Xanthan in the outer water phase, molecular exchange between W_1 and W_2 occurs. However, the exchange rate is that low that it is negligible on the time scale of the NMR experiments. This observation is proofed by pictures made with a CLSM, shown in Figure 5.

Here, a double emulsion with Xanthan in W_2 was recorded before adding additional W_2, containing 15% sugar. The process was observed online while adding the phase. During the disappearance of W_1 it was observed in the CLSM experiments that the W_1-droplets near the interface of the oil droplets to W_2 tend to move fast along the boundary of the oil droplet. In contrast to that, the W_1-droplets in the centre of the oil droplets seem not to be influenced by the sugar in the outer water phase and do not move on a short time scale.

The impression is that the exchange processes start from the interface of the oil droplets to W_2, but no evidence is found that complete W_1-droplets break the oil droplet's interface. A rearrangement of the remaining W_1 droplets toward the droplet's boundary occurs on a longer time scale as the system aims for concentration equilibrium. This process can be observed until the W_1-droplets disappeared completely.

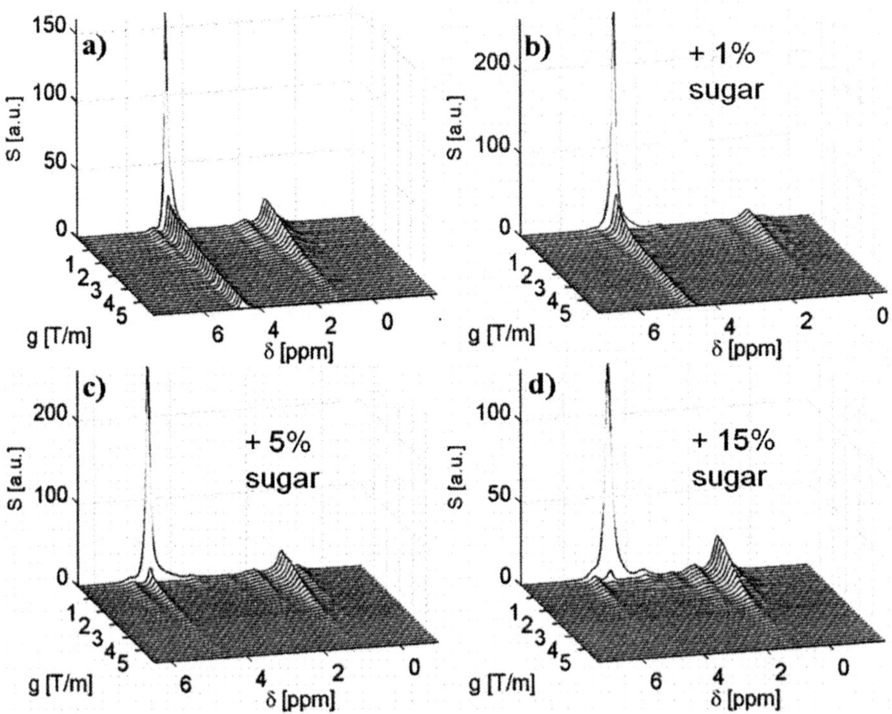

Figure 4 *NMR spectra of a) a WOW double emulsion wit Xanthan in W_2, b) the same emulsion with additional W_2 containing 1% sugar, c) 5% sugar, d) 15% sugar.*

2.2.4 Two Compartment Model. With the knowledge from these experiments, it is possible to separate the NMR signals of W_1 and W_2 in a $W_1/O/W_2$ emulsion with Xanthan in W_2. A two compartment model is applicable. The two water phases can be regarded as independent and completely separate on the experimental time scale on which the exchange is negligible. The total water signal is observed near $g = 0$ T/m. At higher gradient amplitudes, W_2 is negligible or even not measurable. To eliminate the signal of outer water, the measured points over q^2 can be cut at the second gradient step. Discarding the signal near $g = 0$ T/m, the signal of W_1 can now be processed with the MC-model. Therefore, the inner water phase of double emulsions of the type $W_1/O/W_2$ can be characterised, and its DSD is found to be consistent with other measuring techniques.

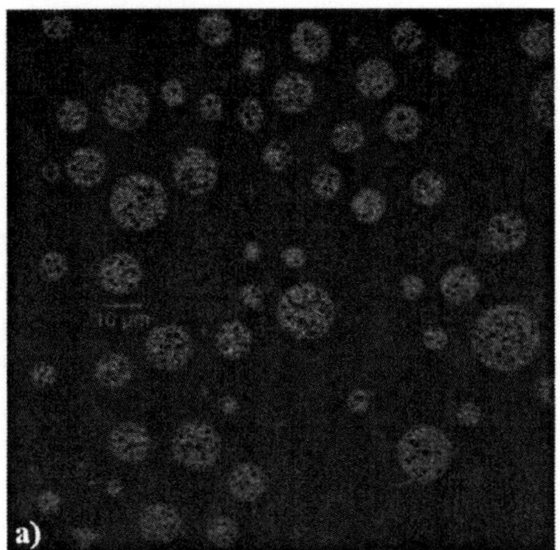
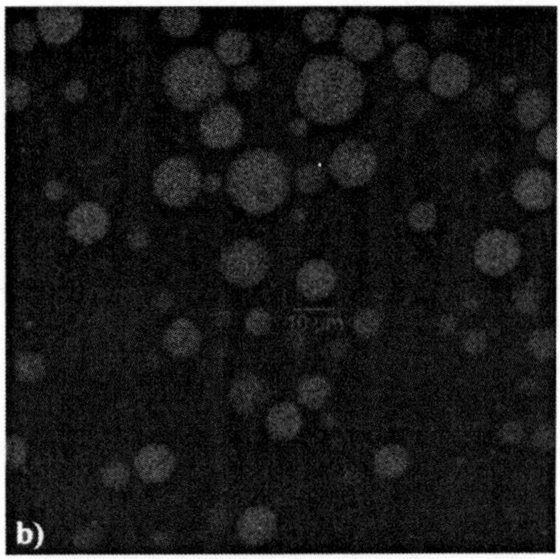

Figure 5 *CLSM pictures of a) a $W_1/O/W_2$ double emulsion with Xanthan in W_2 before adding additional W_2 with 15% sugar. b) a $W_1/O/W_2$ double emulsion with Xanthan in W_2 after adding additional W_2 with 15% sugar.*

3 CONCLUSIONS

PFG-NMR is a suitable method to characterise $W_1/O/W_2$ double emulsions with respect to DSD and molecular exchange phenomena. Due to the molecular exchange between both water phases, a characterisation with the common models as the MC-model is not generally possible. With Xanthan in W_2 of $W_1/O/W_2$ double emulsions, a two compartment model is applicable. The water phases can be regarded as independent. Thereby, the separated signal of the inner water phase can be evaluated by the MC-model applicable and known from investigations of single emulsions. In case of other $W_1/O/W_2$ recipes, the exchange model can successfully be applied, revealing the DSD of the inner emulsions.

Acknowledgements

The 'Shared Research Group 10-2' received financial support by the 'Concept for the future' of Karlsruhe Institute of Technology within the framework of the German Excellence Initiative. Additional support was given the DFG, which are also highly appreciated. We would like to thank K. Sachsenheimer for help during the NMR experiments and Lydia Schütz for emulsion preparation. Special thanks to Azad Emin for help with the CLSM experiments.

References

1 P. T. Callaghan, *Principles of nuclear magnetic resonance microscopy*, Clarendon Press, Oxford, 1991.
2 M. L. Johns and K. G. Hollingsworth, *Progress in NMR Spectroscopy,* 2007, **50**, 51.
3 J. S. Murday, R. M. Cotts, *J. Chem. Phys.*, 1968, **48**, 4938.
4 C. H. Neuman, *J. Chem. Phys.*, 1974, **60**, 4508.
5 J. Pfeuffer, U. Flögel, D. Leibfritz, *NMR Biomed.*, 1998, **11**, 11.

6 J. Pfeuffer, U. Flögel, D. Leibfritz, *NMR Biomed.*, 1998, **11**, 19.

7 W. S. Price, A.V. Barzykin, K. Hayamizu, M. Tachiya, *Biophys. J.*, 1998, **74**, 2259.

8 B. Balinov, O. Weman, J.C. Ravey, *J. Phys. Chem.*, 1994, **98**, 393.

9 J. P. Hindmarsh, J. Su, J. Flanagan, H. Singh, *Langmuir* 2005, **21**, 9076.

10 F. Wolf, L. Hecht, H. P. Schuchmann, E. H. Hardy, G. Guthausen, *Eur. J. Lipid Sci. Technol.*, 2009, **111**, 730.

11 X. Guan, K. Hailu, G. Guthausen, F. Wolf, R. Bernewitz, H. P. Schuchmann, *Eur. J. Lipid Sci. Technol.*, 2010, **112**, 828.

12 G. Muschiolik, I. Scherze, P. Preissler, J. Weiss et al., *Multiple emulsions – preparation and stability*, 13th World Congress of Food Science & Technology "Food is Life", 2006, 17–21.

QUANTIFICATION OF OLIGOSACCHARIDES FROM COMMON BEANS BY HR-MAS NMR

L.M.Lião,[1] E.G. Alves Filho,[1] L.M.A. Silva,[1] R. Choze,[1] G.B. Alcantara,[1] P.Z. Bassinello[2]

[1]Instituto de Química, Universidade Federal de Goiás, CP 131, 74001-970, Goiânia, Brazil.
[2]Embrapa Arroz e Feijão, Rodovia GO-462, km 12 Zona Rural CP 179, 75375-000, Santo Antônio de Goiás, Brazil.

1 INTRODUCTION

Common beans are an important and inexpensive source of protein, dietary fiber, iron, complex carbohydrates, minerals and vitamins for millions of people in the world.[1] Although the protein potential is high, they may present antinutritional factors and other substances harmful to health, such as enzyme protease inhibitors, lectins, antivitamins, tannins, flatulence factors, allergenics, phytates and toxins. Enzyme inhibitors can diminish protein digestibility, and lectins can reduce nutrient absorption, but both have little effect after cooking.[2] Phytic acid can diminish mineral bioavailability.[3] The flatulence factors are caused by the oligosaccharides raffinose, stachyose and verbascose (Figure 1) that, due to the absence of α-galactosidase in humans, are anaerobically fermented by microorganisms to produce carbon dioxide, hydrogen and methane.[4] However, these non-digestible oligosaccharides have been identified as prebiotic agents, i.e. food ingredients potentially beneficial to the health of consumers.[5] Oligosaccharides can be fermented in the colon to produce short-chain fatty acids such as acetic, propionic, and butyric acids, and these compounds have been shown to decrease the incidence of colon cancer induced by azoxymethane.[6]

Consequently to find common bean cultivars that produce these oligosaccharides in lower concentrations, minimizing antinutritional effects but maintaining nutraceutical aspects, could be an important strategy to improve beans consumption. In this context we describe a new method to quantify oligosaccharides in common beans using ^{1}H High Resolution Magic Angle Spinning – Nuclear Magnetic Resonance (^{1}H HR-MAS NMR).

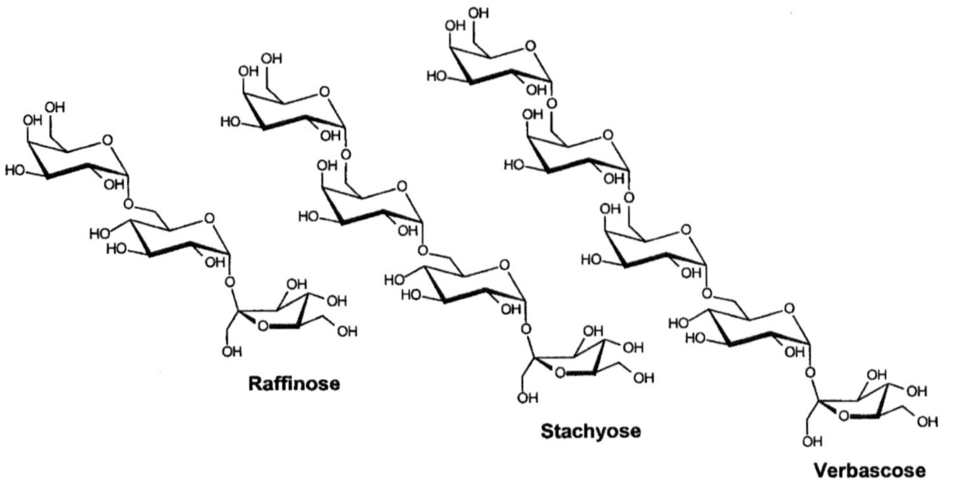

Figure 1   *Raffinose, stachyose and verbascose structures*

HR-MAS NMR technique combines the typical advantages of solid and liquid-state NMR techniques and has recently been reported in literature as an analytical tool in study for solid food product investigations without any pretreatment.[7,8] The method works because of the high degree of internal molecular motion of the materials under study, which has the effect of reducing dipolar couplings, chemical shift anisotropies and magnetic susceptibility effects such that they can be removed by spinning at the magic angle (54.7°) at speeds of about 5 kHz. The aim is to avoid as much as possible the manipulation of samples and to obtain spectra with enough information. Food material is not really solid, there always is a certain amount of movement possible. In this case dipolar interactions are reduced in size and can be removed by rapid spinning only.

2 MATERIAL AND METHODS

2.1   Plant Materials

Five cultivars of different colour groups of common beans (*Phaseolus vulgaris*), BRS MG Tesouro, BRS 7762 Supremo, Jalo Precoce, Pérola and Radiante, acquired under the same conditions and planting season (September/2009) were obtained from the Agriculture Research Station, EMBRAPA – Rice and Beans, Santo Antônio de Goiás, Goiás, Brazil. Only the healthy and homogeneous grains of each cultivar were selected for the analyses.

2.2   Sample Processing

For ^{1}H HR-MAS NMR all the beans were peeled, powdered using liquid nitrogen, and 30 mg suspended in 110 mg of D_2O/TMSP-D4 (sodium-3-trimethylsilylpropionate-2,2,3,3-d4) solution, used as internal standard. For extracts evaluation 100 g of each peeled bean grains were soaked in 600 mL of water during 16 hours in room temperature, and the soaking water was removed. Others 600 mL of water were added and bean grains were submitted to pressure cooking process during the time previously established on Mattson cooker apparatus (25 weighted plungers)[9], as showed on Table 1. The soaking and cooking water extracts were lyophilized, and 20 mg suspended in 600 µL of D_2O/TMSP-D4 for NMR analyses.

Table 1   *Cooking time of bean cultivars established on Mattson cooker apparatus (25 weighted plungers)*

Bean cultivars	Cooking times (min.)*
BRS MG Supremo	30.88 ± 0.52
BRS MG Tesouro	56.21 ± 0.06
Jalo Precoce	27.63 ± 0.63
Pérola	48.94 ± 0.57
Radiante	30.83 ± 0.31

* average of two replicates

2.3   NMR Spectroscopy

One dimensional ^{1}H NMR experiments were recorded at 28 °C using a Bruker Biospin Avance III 500 spectrometer operating at 500.13 MHz (^{1}H), and equipped with a HR-MAS probehead to powered raw (*in natura*) grain analyses and a TBI (Triple Broadband Inverse) probehead to liquid analyses. The analyses were made in triplicate using Composite Pulse Presaturation Sequence (CPPR) to solvent signal suppression. Both spectra were recorded using 128 scans, 64k data points, spectral widths 8012.8 Hz, acquisition time 4.0895 s and relaxation delay (d1) 2 s, sufficient for the proton with the longest relaxation time in the sample. For HR-MAS analyses a 50 µL rotor spinning at the magic angle and 5 kHz was used.

2.4   Calibration Curve

Five solutions containing 1.0, 2.0, 3.0, 4.0 and 5.0 mg, respectively, of raffinose standard (Sigma-Aldrich) diluted at 600 µL of D_2O/TMSP-D4 solution were prepared in order to obtain calibration curve. The solutions were analyzed in TBI probe. The D_2O/TMSP-D4 solution was prepared using 5.6 mg of TMSP-D4 and 5.0 mL of D_2O.

3 RESULTS AND DISCUSSION

Quantitative ^{1}H NMR accounts for the most quantitative NMR applications, because the response of protons is nearly equal regardless of their chemical shifts or coupling to other nuclei with the important exception of exchangeable protons. Most quantitative NMR assay methods are based on using an internal standard, for particular analyses and improvement of the results. In this attempt, TMSP-D4 was used as an internal standard because it is soluble, inert, stable in sample media, and its signal is not in overlapping region. The intensity of the TMSP signal is also important to improve the detection limit and sensitivity.

^{1}H NMR analysis allowed the choice of anomeric signal to be used in the quantification procedure. According to Figure 1, each one of these oligosaccharides has only one glucose unit. Therefore the peak area of the anomeric proton of glucose from oligosaccharides raffinose, stachyose and verbascose related to that of the sharp signal of TMSP-D4 was employed for the quantitative analyses. Quantification was carried out using the known TMSP-D4 concentration related to oligosaccharides peak areas to determine relative mass.

The chemical shift of the anomeric proton of glucose was confirmed by cross link with fructose quaternary carbon in HMBC experiments. Doublet at δ 5.42 was attributed to raffinose anomeric proton and doublet at δ 5.44 to stachyose and verbascose anomerics (Figure 2).

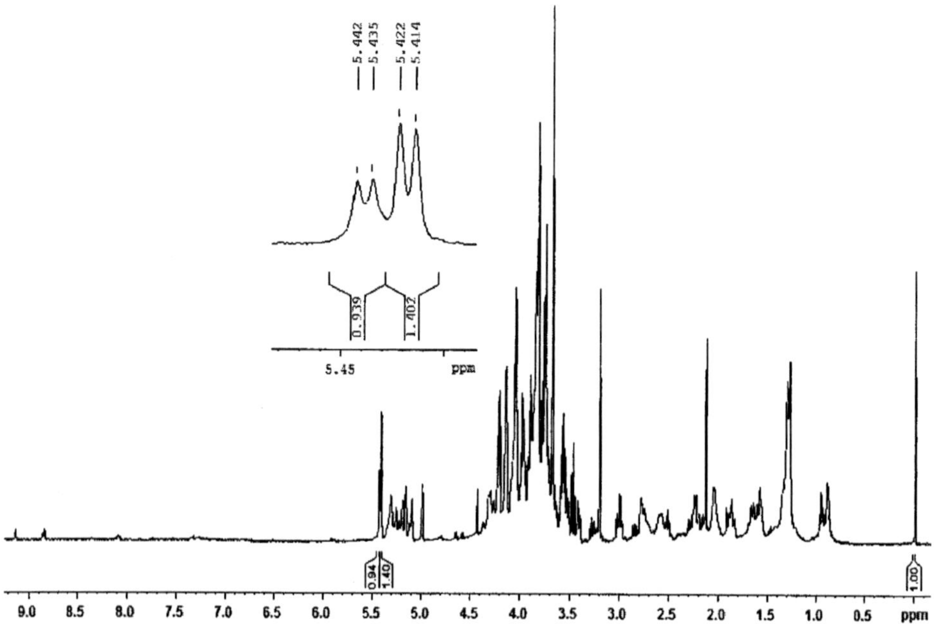

Figure 2   *Jalo Precoce ^{1}H HR-MAS NMR in D$_2$O. Highlighted is anomeric proton region to raffinose (δ 5.42) and mixture of stachyose and verbascose (δ 5.44). The integrals are 1.40 and 0.94, respectively, related to TMSP-D4 (1.00).*

The results of the ^{1}H HR-MAS NMR technique applied to the determination of raffinose, stachyose and verbascose concentrations in common beans are described in Table 2. These results emphasize the BRS 7762 Supremo and Radiante cultivars as well as the most verbascose and stachyose contents. BRS 7762 Supremo cultivar also contains the most quantities of raffinose. On the other hand BRS MG Tesouro cultivar contains the lowest concentration of raffinose, stachyose and verbascose. The contents of oligosaccharides related in this work were higher than related in literature in which values are 0.5 - 4.1% to stachyose and 0.3 – 1.3% to raffinose and verbascose.[10]

Oligosaccharides with very restrict mobility, like in amorphous form, were not detected because HR-MAS technique only permits the detection of molecules that contain some mobility. Therefore this fact was considered in quantification of total oligosaccharides.

Table 2   *Oligosaccharides concentrations in raw grains (in natura), after soaking and water extracts of common beans*

Bean cultivars	Process	Oligosaccharides in 1g of common beans[a]			
		verbascose & stachyose	(%)	raffinose	(%)
BRS 7762 Supremo	Powered raw grains	0.1969 ± 0.0048	19.7	0.2204 ± 0.0056	22.0
	Grain after soaking	0.0710 ± 0.0010	7.1	0.0750 ± 0.0011	7.5
	Soaking water	0.0135 ± 0.0003	1.4	0.0089 ± 0.0002	0.9
	Cooking water	0.0093 ± 0.0002	0.9	0.0047 ± 0.0001	0.5
BRS MG Tesouro	Powered raw grains	0.0960 ± 0.0022	9.6	0.1110 ± 0.0027	11.1
	Grain after soaking	0.0640 ± 0.0009	6.4	0.0730 ± 0.0010	7.3
	Soaking water	0.0035 ± 0.0001	0.4	0.0058 ± 0.0002	0.6
	Cooking water	0.0040 ± 0.0001	0.4	0.0076 ± 0.0002	0.8
Jalo Precoce	Powered raw grains	0.1418 ± 0.0035	14.2	0.1253± 0.0032	12.5
	Grain after soaking	0.0470 ± 0.0009	4.7	0.0300 ± 0.0008	3.0
	Soaking water	0.0144 ± 0.0003	1.4	0.0052 ± 0.0002	0.5
	Cooking water	0.0101 ± 0.0003	1.0	0.0020 ± 0.0001	0.2
Pérola	Powered raw grains	0.1012 ± 0.0027	10.1	0.1117 ± 0.0026	11.2
	Grain after soaking	0.0520 ± 0.0009	5.2	0.0560 ± 0.0009	5.6
	Soaking water	0.0071 ± 0.0003	0.7	0.0060 ± 0.0003	0.6
	Cooking water	0.0080 ± 0.0003	0.8	0.0072 ± 0.0003	0.7
Radiante	Powered raw grains	0.2146 ± 0.0052	21.5	0.1180 ± 0.0037	11.8
	Grain after soaking	0.0650 ± 0.0010	6.5	0.0520 ± 0.0009	5.2
	Soaking water	0.0103 ± 0.0003	1.0	0.0016 ± 0.0001	0.2
	Cooking water	0.0097 ± 0.0003	1.0	0.0029 ± 0.0001	0.3

[a]Oligosaccharides in amorphous form were not quantified

3.1 Soaking and Cooking Effects

According to Khokhar and Chauhan,[11] soaking and cooking processes are known to reduce the anti-nutritional factors, improving the nutritional value of legumes. To quantify the efficiency of these processes in oligosaccharides removal by ^{1}H NMR, one group of bean cultivars were soaked in water during 16 hours and after this submitted to pressure cooking during time presented in Table 1. The quantities of raffinose, stachyose and verbascose in water extracts and grains are also reported in Table 2.

The average reduction in soaking was 56.4% for raffinose and 56.5% to stachyose and verbascose. The oligosaccharides quantities observed in water extract from this process compared with reduction in bean grains suggested a decomposition of these oligosaccharides. This fact was supported by signals of galactosides and glucosides observed in ^{1}H NMR spectra, caused by hydrolysis of oligosaccharides. These signals were not observed in raw beans spectra obtained by HR-MAS (Figure 3).

The decomposition of oligosaccharides was also observed in cooking process. The HR-MAS NMR spectra of cooked grains showed overlapped signals in the anomeric proton region being impossible to quantify raffinose, stachyose and verbascose anomeric proton by integration (Figure 3). Onigbinde and Akinyele[12] proposed that raffinose and stachyose content in grains of 20 varieties of cowpeas (*Vigna unguiculata*) decreased after cooking and they also observed an increase in sucrose level after cooking. In cooked beans water extracts the oligosaccharides content average was 10.7% to raffinose and 13.6% for stachyose and verbascose.

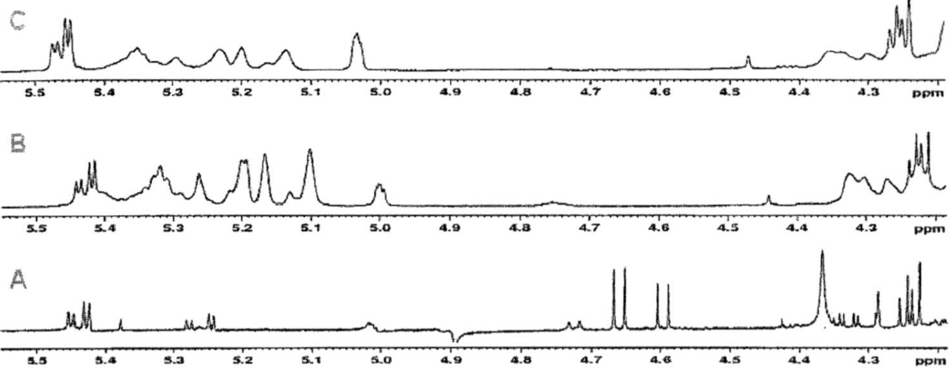

Figure 3   *Galactosides and glucosides observed in ^{1}H NMR spectra, caused by hydrolysis of oligosaccharides after soaking (A) and overlapping signals after cooking processes (B). ^{1}H HR-MAS NMR to raw (in natura) common beans (C)*

3.2   Calibration Curve

The linear relationship between the concentration of raffinose standard and anomeric proton integration was used to confirm the proposed methodology to quantify raffinose, stachyose and verbascose on common beans by ^{1}H NMR. A series of raffinose standard solutions on range from 1.0 to 5.0 mg in D_2O/TMSP-D4 solution was prepared. The anomeric proton integration plotted *versus* solution concentration produced a linear correlation coefficient of 0.9946 on the basis of the peak δ 5.42. The refereed peak was chosen because its signal is not in the overlapping region.

4 CONCLUSION

The ^{1}H HR-MAS NMR analyses pointed out the BRS MG Tesouro common bean variety as the on with the lower raffinose, stachyose and verbascose content. On the other hand this cultivar presented the longest cooking time, 56 minutes, increasing the difficulty in its commercialization. This fact is in agreement with the aim of Embrapa Rice and Beans to produce commercially accepted bean varieties which contain as much as possible lower quantities of these oligosaccharides. These analyses also demonstrated that soaking and cooking are efficient processes to reduce oligosaccharides content in bean grains.

This study also emphasizes the remarkable advantage in using the HR-MAS NMR technique for the determination of raffinose, stachyose and verbascose content in common bean grains. The oligosaccharides measurement is highly simplified because it does not require any pretreatment of the sample apart from the addition of a small amount of D_2O necessary to produce homogeneous dough and a field frequency lock. Moreover, due to the high concentration of the sample, measurement time in HR-MAS NMR is very short. On the contrary, the water extracts evaluation showed the presence of decomposed products distorting oligosaccharides content results, what demonstrates the disadvantage of quantification process by using a solution NMR in the methodology.

Acknowledgements

The authors thank the CNPq, MCT/FINEP/CT-INFRA and FUNAPE for their financial support and CAPES for the fellowship provided to E.G. Alves Filho.

References

1   C. Reyes-Moreno and O. Paredes-López, *Crit. Rev. Food Sci. Nutr.*, 1993, **33**, 227.
2   F.M. Lajolo and M.I. Genovese, *J. Agric. Food Chem.*, 2002, **50**, 6592.
3   A.S. Sandberg, *Br. J. Nutr.*, 2002, **88**, S281.
4   K.R. Price, J. Lewis, G.M. Wyatt and G.R. Fenwick, *Nahrung,* 1988, **32**, 609.
5   J. van Loo, J. Cummings, N. Delzenne, H. Englyst, A. Franck, M. Hopkins, N. Kok, G. Macfarlane, D. Newton, M. Quigley, M. Roberfroid, T. van Vliet, and E. van den Heuvel, *Br. J. Nutr.,* 1999, **81**, 121.
6   L. Hangen and M.R. Bennik, *Nutr. Cancer*, 2002, **44**, 60.
7   M.A. Brescia, G. Di Martino, C. Fares, N. Di Fonzo, C. Platani, S. Chelli, F. Reniero and A. Sacco, *Cereal Chem.*, 2002, **79**, 238.
8   L.M. Lião, R. Choze, P.P.A. Cavalcante, S.C. Santos, P.H. Ferri and A.G. Ferreira, *Quim. Nova*, 2010, **33**, 634.
9   J.R. Proctor, B.M. Watts, *Can. Inst. Food Sci. Technol. J.*, 1987, **20**, 9.
10 N.R. Reddy, M.D. Pierson, S.K. Sathe and D. K. Salunkhe, *Food Chem.,* 1984, **13**, 25.
11 S. Khokhar and B.M. Chauhan, *J. Food Sci.*, 1986, **51**, 591.
12 A. O. Onigbinde and I.O. Akinyele, *J. Food Sci.*, 1983, **48**, 1250.

STUDY OF THE PROTEOLYTIC AND LIPOLYTIC PROCESSES IN MANCHEGO CHEESE BY NMR

M. Moreno,[a] A. Moreno,[a] M. V. Gómez,[b] J. M. Povéda[b] and L. Cabezas.[c]

[a] Department of Química Inórganica, Orgánica y Bioquímica, University of Castilla-La Mancha. Facultad de Química, Avenida Camilo José Cela 10, 13071 Ciudad Real, Spain. Email: Andres.Moreno@uclm.es
[b] Instituto Regional de Investigación Científica Aplicada. University of Castilla La Mancha. Avenida Camilo José Cela s/n, 13071 Ciudad Real, Spain.
[c] Department of Bromatología y Tecnología de Alimentos, University of Córdoba. Campus Rabanales, 14014 Córdoba, Spain.

1 INTRODUCTION

Cheese is a complex matrix which can change very much depending upon its origin, conditions of the manufacturing processing, microbial flora, enzyme activities, freshness, or ripening time. Lipolysis plays an essential role in the sensory properties of cheese; some free fatty acids have been shown to contribute to the aroma and the flavour of cheese. Moreover, conjugated linoleic acids (CLA) are really interesting for its antitumoral, immunomodulating and antidiabetic activities. On the other hand, proteolysis is one of the major biochemical events which takes place during cheese ripening and its products, free amino acids and peptides, have a considerable influence on the sensory characteristics of the cheese.

Several techniques have been used to monitor and quantify proteolysis and lipolysis process in cheese during ripening. Actually, NMR spectroscopy is emerging as an alternative analytical tool in a number of applied fields, including Food Science.[1,2] It is possible to identify and quantify many compounds in a complex mixture simultaneously and nondestructively.

2 METHOD AND RESULTS

2.1 Sample Preparation

Samples of Manchego cheese made from raw milk and from pasteurised milk and aged for 3, 6 and 9 months were analysed in this study by NMR spectroscopy. pH values were measured for all samples and resulted between 4.8 and 5.8. For this reason we did not add any buffer, also to preserve the compound ratio in the water solution.

For the organic solvent extraction, an amount (0.5 g) of ground sample was weighed directly in a test tube and extracted with 1 mL of deuterium chloroform, vortexing for 15 min. The extract was centrifugated at 14000 rpm for 15 min at 4 °C and the supernatant was filtered using a Millipore filter PTFE 0.45 µm. For all NMR spectra, 0,5 mL of extract were transferred into 5 mm NMR tubes.

In the case of the ^{31}P-NMR, we use a solution, which consists of solving 0.6 mg of chromium acetylacetonate (III), Cr(acac)$_3$ in 10 mL of a mixture of pyridine and CDCl$_3$ solvents (1.6:1 volume ratio), and it was protected from moisture with molecular sieves.[3] The samples were prepared by adding 20 µL of the solution and 20 µL of TMDP (2-chloro-4,4,5,5-tetramethyl-1,3,2-dioxaphospholane). The mixture was left to react for half an hour (Scheme 1), and then was used to acquire the ^{31}P-NMR spectra.[3] Phenol was used as internal standard in order to quantify the different compounds after NMR analysis.

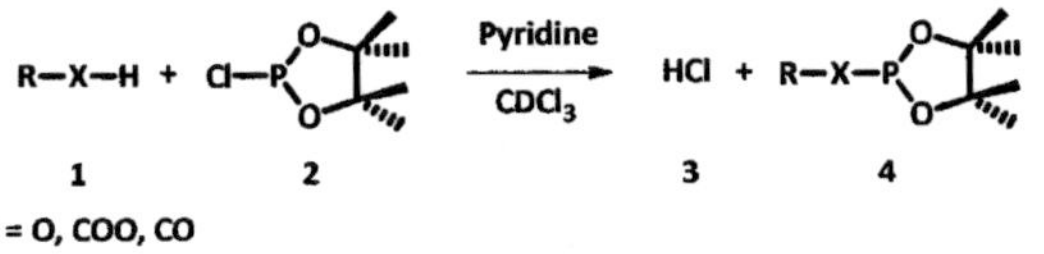

Scheme 1

On the other hand, for water-soluble metabolites, portions of samples (0.5 g) were weighed directly in a test tube and extracted with 1 mL of Milli-Q water, vortexing for 15 min. The extracts were centrifugated at 14000 rpm for 15 min at 4 °C and the supernatants were filtered using a Millipore filter Nylon 0.45 µm to remove proteins and peptides. All samples were lyophilized. For all NMR spectra, the lyophilized sample was dissolved in 0,5 mL of D$_2$O and was transferred into 5 mm NMR tubes. DMS (dimethylsulfone) was used as internal standard in order to quantify the different compounds.

2.2 NMR spectroscopy

All NMR spectra were recorded at 25 °C, on a Varian Inova 500 MHz spectrometer for ^{1}H and ^{13}C-NMR, and a Varian Gemini 400 MHz spectrometer for ^{31}P-NMR. Longitudinal relaxation time T1 of all signals of interest was determined using the inversion-recovery sequence. Peak identification was confirmed through the addition of pure analyte and through 2D spectra: COSY, HSQC, HMQC, TOCSY and INADEQUATE experiments.

2.3 qNMR

The determination of ratios of components in drug, natural products or synthesized compound mixtures is easy to achieve by qNMR using the integrals.[5] It is possible to use a qNMR absolute method for quantitative analysis of the content or concentration: The main component P_X can be calculated directly using a standard of known content P_{Std} (Eq. 1)

$$P_X = \frac{I_X}{I_{Std}} \cdot \frac{N_{Std}}{N_X} \cdot \frac{M_X}{M_{Std}} \cdot \frac{m_{Std}}{m} \cdot P_{Std} \qquad \textbf{Eq. 1}$$

with M_X and M_{Std} being the molecular weights of analyte and standard, m and m_{St} the weights of X and the standard in the sample, N_X and N_{Std} the number of contributing nuclei of the signals of X and standard considered, and P_X and P_{Std} the concentration of analyte and standard, respectively.

2.4 Identification and quantitative determination of Organic-Soluble Metabolites

A fast and reproducible extraction of the organic fraction of Manchego cheese, has allowed us to study tri-, di-, monoglycerides, sterols and free fatty acids. Focusing on the ripening time, samples were analyzed by ^{1}H-NMR (Figure 1), ^{13}C-NMR (Figure 2) and ^{31}P-NMR (Figure 3) upon derivatization of hydroxyl and carboxyl groups with a phosphorous reagent (Scheme 1) to identify the minor compounds. The main chemical shifts of the interesting compounds are given (Table 1, 2 and 3).

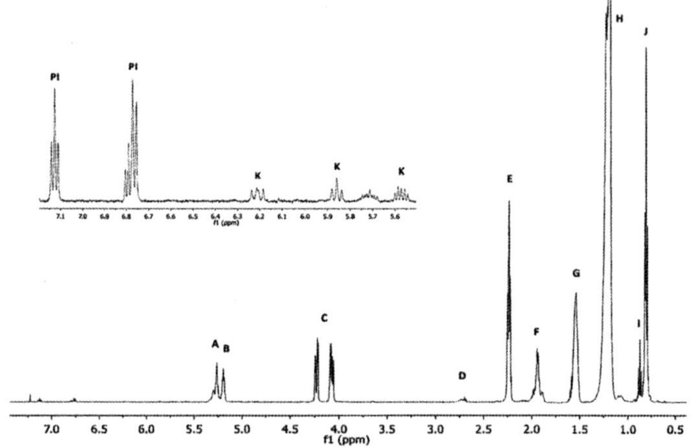

Figure 1 *^{1}H-NMR spectrum of organic extract of cheese*

Table 1 *Main chemical shifts (ppm) from ^{1}H-NMR spectrum*

Signal	δ (ppm)	^{1}H	Compound
PI	7.25, 6.77	-CH=CH-	Phenol
K	6.22, 5.88, 5.56	-CH=CH-CH=CH-	CLA
A	5.29	-CH=CH-	Insaturated Fatty Acids
B	5.15	-CH-O-COR	Triglycerides
C	4.19	-CH$_2$-O-COR	Triglycerides
D	2.76	-CH=CH-CH$_2$-CH=CH-	Linolein and Linolenin
E	2.20	-CH$_2$-COOH	Free Fatty Acids
F	2.02	-CH$_2$-CH=CH-	Insaturated Fatty Acids
G	1.60	-CH$_2$-CH$_2$-COOH	Free Fatty Acids
H	1.20	-(CH$_2$)$_n$-	Fatty Acids
I	0.95	-CH=CH-CH$_2$-CH$_3$	Linolenin
J	0.85	-CH2-CH2-CH$_2$-CH$_3$	Fatty Acids

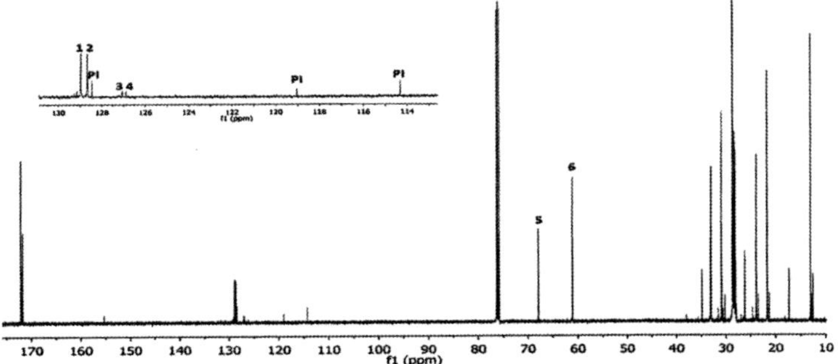

Figure 2 *^{13}C-NMR spectrum of organic extract of cheese*

Table 2 *Main chemical shifts (ppm) from ^{13}C-NMR spectrum*

Signal	δ (ppm)	^{13}C	Compound
PI	129.4, 119.8, 115,3	-CH=CH-	Phenol
1	130.0	-CH$_2$-CH=CH-CH$_2$-	Olein (C-10)
2	129.7	-CH$_2$-CH=CH-CH$_2$-	Olein (C-9)
3	128.1	-CH=CH-CH$_2$-CH=CH-	Linolein (C-10)
4	127.9	-CH=CH-CH$_2$-CH=CH-	Linolein (C-12)
5	67.5	-CH-O-COR	Trigliycerides (sn-2)
6	63.4	-CH$_2$-O-COR	Trigliycerides (sn-1,3)

^{1}H, ^{13}C-NMR and homonuclear correlation experiments (gCOSY, NOESY, TOCSY), heteronuclear correlation experiments (gHSQC, gHMBC) allows to differ long-chain fatty acids (saturated and unsaturated, like oleic and linoleic acids), such as free fatty acid or as esterified with glycerol, ie, triglycerides. Their distribution in the backbone of glycerol could be studied using the INADEQUATE pulse sequence and gHMQC-TOCSY tridimensional experiment.

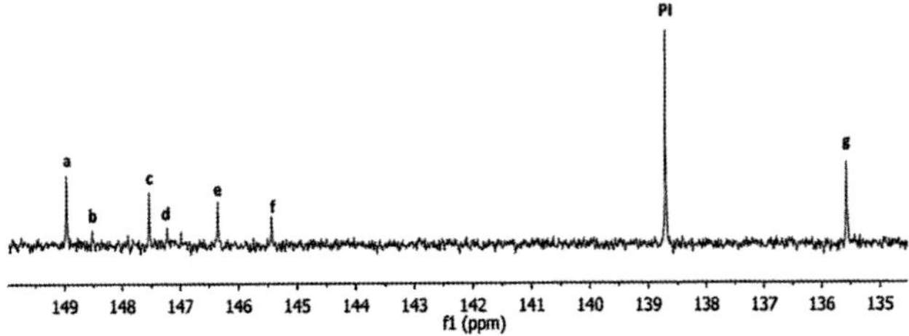

Figure 3 *^{31}P-NMR spectrum of organic extract of cheese*

Table 3 *Main chemical shifts (ppm) from ^{31}P-NMR spectrum*

Signal	δ (ppm)	^{31}P	Compound
PI	138.7	-C-O-P-O$_2$-	Phenol
a	149.0	-CH$_2$-O-P-O$_2$-	1,2-digllyceride
b	148.5	-CH$_2$-O-P-O$_2$-	1-monoglyceride (sn-1,3)
c	147.5	-CH-O-P-O$_2$-	2-monoglyceride
d	147.2	-CH-O-P-O$_2$-	1-monoglyceride (sn-2)
e	146.4	-CH-O-P-O$_2$-	1,3-diglyceride
f	145.4	-CH-O-P-O$_2$-	Sterols (cholesterol)
g	135.5	-CH$_2$-CO-O-P-O$_2$-	Free Fatty Acids

The results have allowed us to quantify different compounds present in cheese in order to study the lypolitic process. The graphic (Figure 4) shows, as expected, that triglyceride content decrease while di- and monoglycerides, as well as, free fatty acids content, increase with the ripening time.

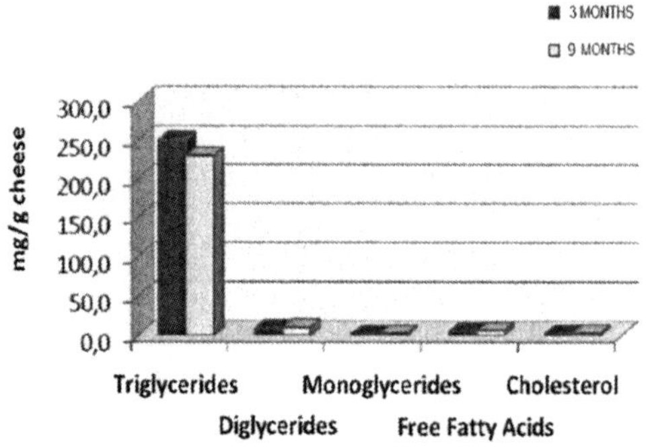

Figure 4 *Lipid profile of a cheese organic extract quantify by qNMR.*

2.5 Identification and quantitative determination of Water-Soluble Metabolites

Water-soluble metabolites, specifically, free aminoacids content of Manchego cheese has been related to ripening (3, 6 and 9 months) by the use of high resolution ^{13}C-NMR spectroscopy.

The complete spectrum of a cheese extract acquired with a single-pulse sequence is reported (Figure 5) The spectrum shows strong signals corresponding to the carbons of lactate and weaker signals which are free aminoacids present in the extract. Peak identification was confirmed through the addition of pure analyte and using solutions of mixtures of different aminoacids, to identify each of them in a complex matrix.

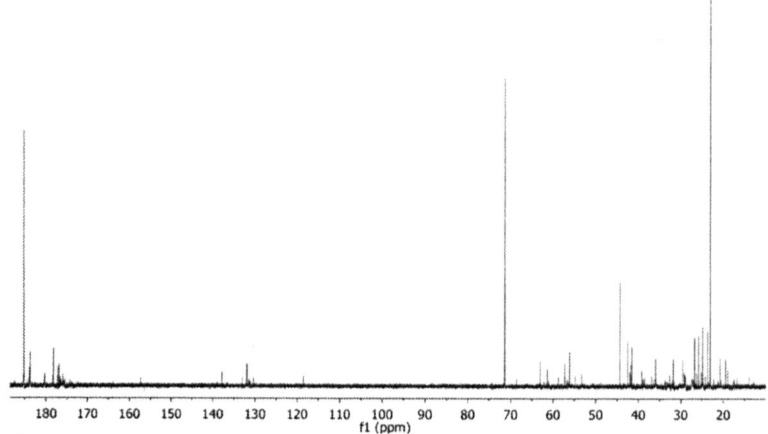

Figure 5 *^{1}H-NMR spectrum of aqueous extract of cheese*

To quantify the different free aminoacid content, we exclude the signals overlap of other compounds. Using the integrals of the chosen signals and the equation 1, we can quantify the free aminoacids present in different cheeses. With the results obtained for cheeses with 3, 6 and 9 months of ripening, in the graphic (Figure 6) is possible to understand the lypolitic process. The free aminoacids content increase with the ripening time, just because protein in cheese is broken down from casein to low molecular peptides and amino acids by various kinds of proteolytic enzymes.[6]

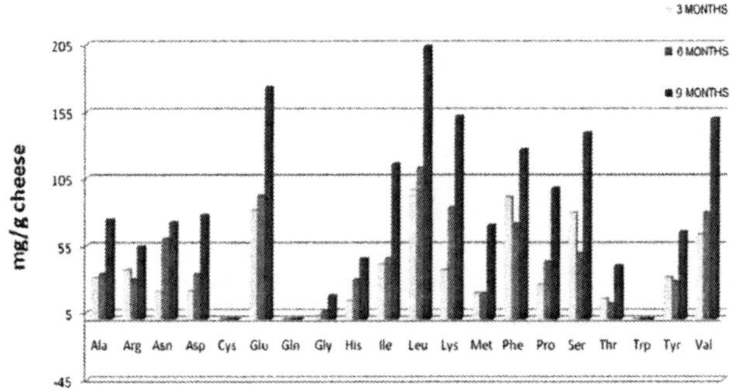

Figure 6 *Free aminoacid profile of a cheese aqueous extract quantify by qNMR.*

3 CONCLUSION

A fast and reproducible extraction of the organic fraction of Manchego cheese has allowed us to study lipid content, specially, conjugated linoleic acids which are really interesting for its antitumoral, immunomodulating and antidiabetic activities. Focusing on the ripening time, samples have been analyzed by ^{1}H, ^{13}C and ^{31}P-NMR. On the other hand, the water-soluble metabolites, specifically, aminoacids content of Manchego cheese have been studied by the use of ^{13}C-NMR spectroscopy. Therefore, it is possible to identify and quantify the different compounds present in cheese without a tedious process of extraction and derivatization.

Lipid profile (Figure 4) and amino acid profile (Figure 6) confirm the significant variations, as expected in ripening, giving an index of the lipolytic and proteolytic processes, respectively.

NMR spectroscopy allows establishing a quantification methodology of each of the compounds found in different cheeses, using an internal standard and/or a derivatizating agent.

References

1. P. S. Belton, I. J. Colquhoun and B. P. Hills, *Annu. Rep. NMR Spectrosc.* 1993, **26**, 1.
2. E. Alberti, P. S. Belton and A. M. Gil, *Annu. Rep. NMR Spectrosc.* 2002, **47**, 109.
3. E. Hatzakis, G. Dagounakis, A. Agiomyrgianaki and P. Dais, *Food Chem.* 2010, **122** (1), 346.
4. P. Dais and A. Spyros, *Magn. Reson. Chem.* 2007, **45**, 367.
5. U. Holzgrabe, *Prog. Nucl. Mag. Res. Sp.* 2010, **57**, 229.
6. J. M. Poveda; L. Cabezas; P. L. H. McSweeney, *Food Chem.* 2004, **84**, 213.

QUALITY MARKERS OF RED WINES FROM SPANISH REGION OF CASTILLA-LA MANCHA USING NUCLEAR MAGNETIC RESONANCE.

A. Moreno,[a] L. F. Labrador,[a] M. Moreno,[a] M. S. Pérez,[b] M. A. González,[b] E. M. Sánchez-Palomo,[b] J. M. Lemus.[b]

[a] Department of Inorganic, Organic Chemistry and Biochemistry, University of Castilla-La Mancha, Avenida Camilo José Cela 10, 13071 Ciudad Real, Spain.
Email: Andres.Moreno@uclm.es
[b] Department of Química Analítica y Tecnología de los Alimentos. University of Castilla-La Mancha. Avenida Camilo José Cela, s/n. 13071. Ciudad Real.

1 INTRODUCTION

Wine is a relatively complex biochemical system primarily consisting of water, ethanol, organic acids, and carbohydrates. In addition to the main constituents a range of aromatic compounds, polyphenols and colorants are present in small quantities.[1] They are important ingredients with respect to quality characteristics such as taste and color. In recent years the composition of wine has been the subject of increased focus because of demands from consumers regarding the quality and authenticity. In fact, quality wines are often produced in restricted areas defined as denomination of controlled origen (DOC). "La Mancha" is the most typical DOC of the Castilla-La Mancha region in Spain. Actually enology and viticulture are two very important sector of the Mancha economy, one of the biggest producers and exporters of wine among the Spanish region.

Nuclear magnetic resonance (NMR) spectroscopy has gained an outstanding role in the characterization and quality control of food.[2,3] NMR has been used mainly as a qualitative tool for the rapid determination of food components. Less frequent is its application in quantitative analysis.[4-6] ^{1}H NMR spectroscopy is a strong tool for assessing additional wine constituents, such as aromatics, carbohydrates and acids can be detected and quantified by this method.[7] NMR has the advantages of being nondestructive and intrinsically more information-rich with respect to the determination of molecular structures, especially in complex mixture analyses. Moreover, in the case of wines,[8-10] the study of their differentiation according to vine variety, geographical origin, and vintage is important for authenticity assessment and in studies of possible adulteration. NMR characterization of the organic matrix of monovarietal selected wines from different areas might be important in determining the possible variations related with the pedological and geological substrata of vineyards.

The aim of this work is to use NMR to characterize selected commercial red wines produced in the DOC "La Mancha", a region in the middle of Spain, and ,in the future ,to examine the possibility of differentiating wines from the same grape variety (Cencibel) but produced in different areas. No NMR study on wines from "La Mancha" has been carried out until now.

2 METHOD AND RESULTS

2.1 Sample Preparation and NMR Measurements

Eleven red wine samples were obtained by different vine cultivars in the region of Castilla-La Mancha. All the samples used in this work were monovarietal Tempranillo or Cencibel DOC red wines. In order to prepare the samples, 5 ml of each wine was evaporated under vacuum to remove the ethanol and water, with the aim of reducing the residual water signal in ^{1}H NMR spectra. Finally, the samples were dissolved in D_2O, and the NMR analyses were carried out. The pH of these solutions was always equal to 4.0. All NMR measurements were performed in a 5-mm tube and Varian Inova spectrometer, operating at 500 MHz for ^{1}H at 25 °C. Sodium 3-(trimethylsilyl) propionate d_4 (TSP 98%) was used as an internal chemical shift reference. All samples were prepared and the spectra were recorded three times. For quantitative propose a 90° pulse, a relaxation delay of 25s, 8 scans and dimethylsulphone DMS as addition standard (3.18 ppm and T_1 similar to the ^{1}H of different components to quantify).

2.2 Identification of Wine Components

Some important components of these wines were identified by the assignments of their ^{1}H and ^{13}C resonances using one-dimensional (1D) and two-dimensional (2D) homonuclear and heteronuclear NMR experiments. A few of these components were also quantified.

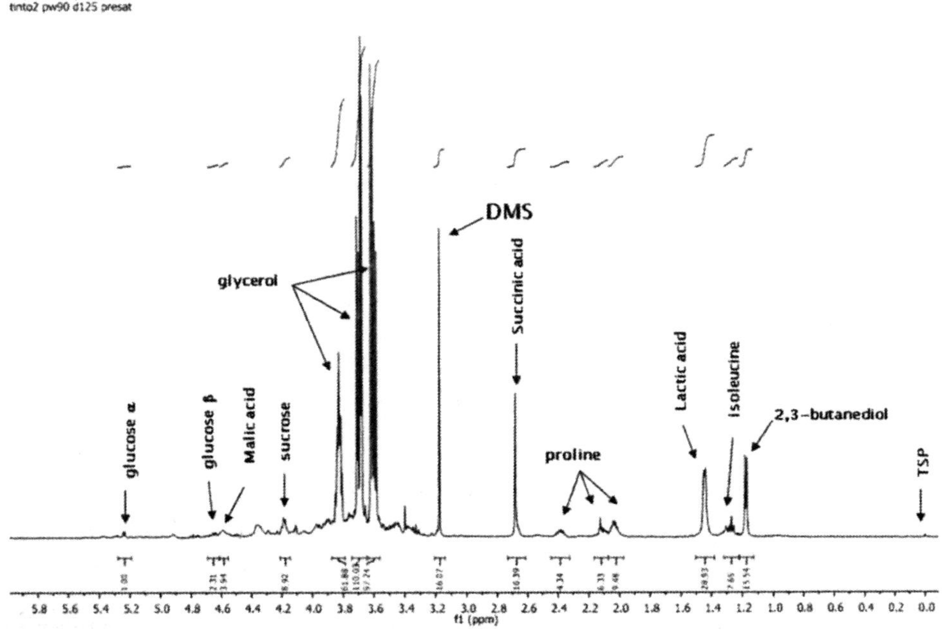

Figure 1 *Representative ^{1}H-NMR wine spectra with water presaturation.*

After the identification of the NMR signals (Figure 1), we make several mixtures of the main components at different concentrations using TSP and DMS as standards for the quantitative analysis.

2.3   Quantitative Calculations

Using integral areas of each artificial mixture or wine spectra, calculations of the quantity of main components may be resolve[4] by two different ways (Eq. 1 and 2):

$$\% \text{ compound}(wt) = \frac{\left. I_{compound} \middle/ N_{compound} \right.}{\sum_{i=1}^{m} \left. I_i \middle/ N_i \right.} \cdot 100 \tag{1}$$

Equation 1 *For calculates the compounds' percentage.*

$$\text{Compound}(\text{mg}) = \frac{I_{compound}}{I_{standard}} \cdot n\,(\text{mmol standard}) \cdot M_w\,(\text{compound}) \tag{2}$$

Equation 2 *For calculates the amount of each interesting compound (mg).*

The accuracy, precision and response linearity of such a procedure were validated by several experiments carried out on pure compounds and artificial mixtures.

2. 4 Artificial Mixtures

We recorded the spectra of different mixtures consisting of glucose, glycerol, succinic acid, proline, lactic acid and 2,3-butanediol, and the mass of each component was calculated by the integral values of the signals showed in Table 1 and according to Eq. 2.

Table 1 *Selected integral areas for the components of artificial mixtures.*

components	group	$\delta\,^1H$ (ppm)	multiplicity	n° hydrogen
Glucose,(α y β)	anomeric	5,23 y 4,65	d	1H
Glycerol	CH_2	3,65	dd	2H
Succinic acid	CH_2	2.61	s	4H
Proline	CH_2	2.34	m	1H
Lactic acid	CH_3	1.33	d	3H
2,3-Butanediol	CH_3	1,15	d	6H

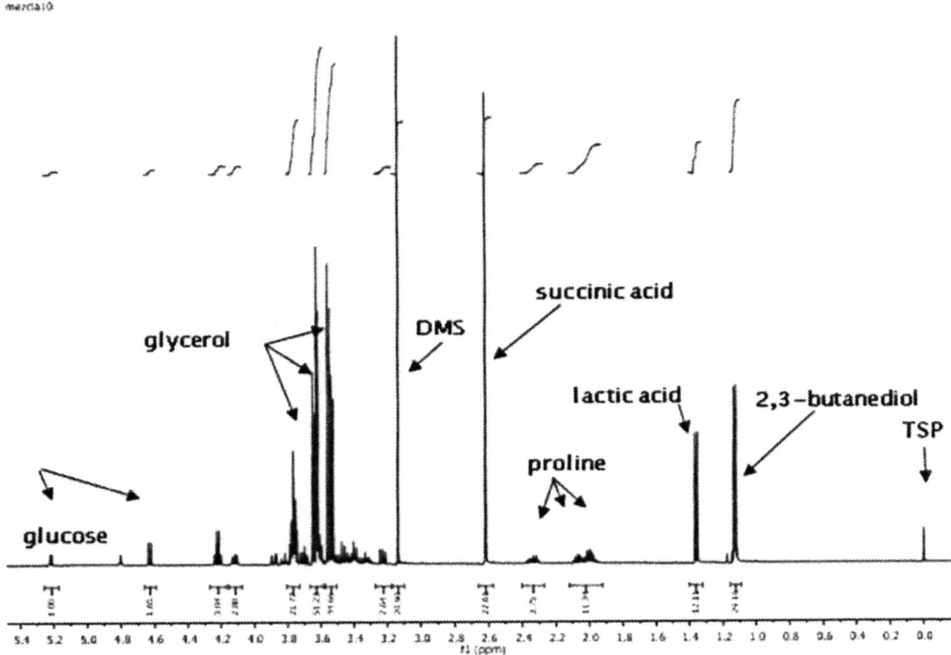

Figure 2 *^{1}H-NMR spectrum of an artificial mixture of main components present in wine.*

To express the weighed of each components as a function of the amounts ones measured by ^{1}H NMR, we drew calibration lines for the six compounds (Figure 3). The response linearity of the experimental procedure was clear from the linear determination factor, R^2, which ranged from 0.896 to 0.998.

2.5 Application to the quantification of wine components.

The results obtained with the pure compounds and the mixture indicated that quantitative determination of wine components can be carried out using spectra acquired with the selected parameters. Therefore, we applied the same procedure to the different wine samples from "La Mancha" DOC. The mass percentage of each component was calculated according to Eq. 1 and the subsequent amount of each component was calculated according to Eq. 2.

The percentages of each one of the components, by the milligrams of extract from 5 mL of wine used in the pre-concentration step are represented in Figure 4.

Finally, it is possible to calculate the milligrams of each compound by their integral values from the integral of DMS, and after the correlation regression, we can indicate the calculated data in mg/L (Table 2 and Figure 5).

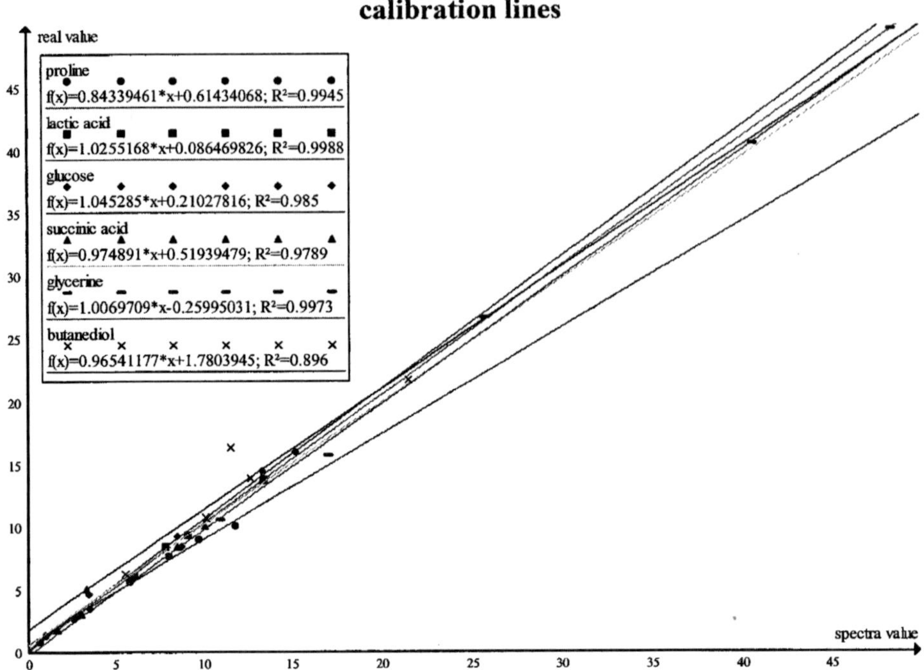

Figure 3 *Calibration lines of proline, lactic acid, glucose, succinic acid, glycerol and 2,3-butanediol. Real value: weighed amount (mg), spectra value: calculated amount (mg) using ^{1}H-NMR spectra according to Eq. 2.*

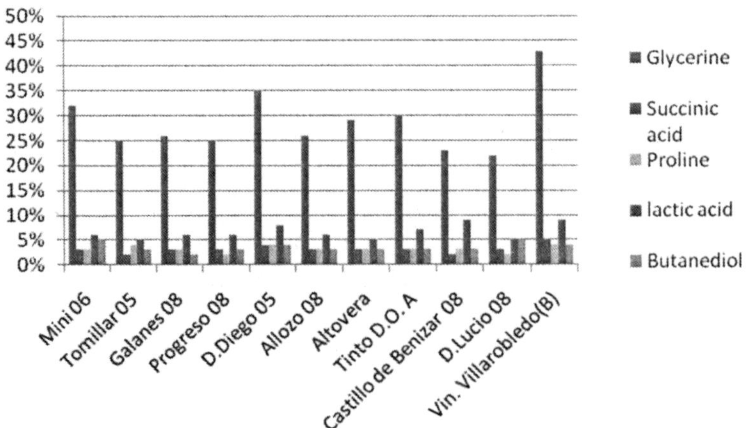

Figure 4 *Graphical observation of the results in different wines and the comparison between the principal components.*

Table 2 *Concentration (expressed in mg/L) determinated by NMR of the wine components of the samples studied in this work from Castilla-La Mancha Region.*

Wine (mg/l)	Glycerine	Succinic Acid	Proline	Lactic Acid	Butanediol
Mini 06	10722	976	1120	1890	1730
Tomillar 05	9256	978	1572	1900	1142
Galanes 08	9580	1084	934	2346	874
El Progreso 08	9488	1060	832	2092	1154
D.Diego 05	11244	1180	1252	2634	1284
Allozo 08	10680	1108	1296	2520	1140
Altovela	10128	1168	1094	1618	1054
Tinto DO. (A)	11270	1244	1072	2536	1042
Castillo 08	7822	778	1144	3198	982
D.Lucio 08	8650	1120	890	2104	1978
Vin. Villarrobledo (B)	14912	1696	1304	3096	1416

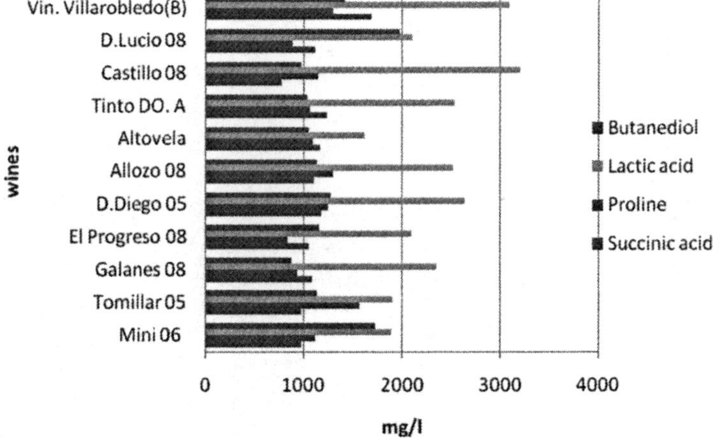

Figure 5 *Graphical observation of the results in different wines and the comparison between the principal components (glycerol not included).*

3 CONCLUSION

This work describes the first preliminary ^{1}H NMR study of samples of wines from DOC "La Mancha" using NMR spectroscopic analysis. The global analysis of metabolites in wine could provide useful information on different quality markers of red wine from Castilla-La Mancha region. The NMR characterization of wines from "La Mancha" that we have presented here is based only on a few well-selected chemical compounds, namely, the organic acids succinic and lactic, the alcohol glycerol and 2,3-butanediol, and the amino acid proline, of which some NMR signals can be easily identified and quantified.

This could find an important application in authenticity studies of wines. In the near future we are considering extending the analysis to a wider selection of Cencibel or Tempranillo wines from other regions. This extended study could be demonstrating the possibility of NMR-based metabolomic research to characterize wine quality and applied

                                    Magnetic Resonance in Food Science

fermentation methods and product origin. Thus a further work is being carried out to widen the scope and quality of wine analysis and characterization.

References

1. R. Clarke and J. Bakker, *Wine Chemistry and Flavour*; Blackwell Publishing Ltd.: Oxford, United Kingdom, 2004; pp 66-116.
2. P. S. Belton, I. J. Colquhoun and B. P. Hills, *Annu. Rep. NMR Spectrosc.* 1993, **26**, 1-53.
3. E. Alberti, P. S. Belton and A. M. Gil, *Annu. Rep. NMR Spectrosc.* 2002, **47**, 109-148.
4. U. Holzgrabe, *Prog. Nucl. Mag. Res. Sp.* 2010, **57**, 229-240.
5. A. Caligiani, D. Acquotti, G. Palla and V. Bocchi, *Anal. Chim. Acta* 2007, **585**, 110–119.
6. P. Petrakis, I. Touris, M. Liouni, M. Zervou, I. Kyrikou, R. Kokkinofta, C. H. Theocaris and T. M. Mavromoustakos, *J. Agric. Food Chem.* 2005, **53**, 5293–5303.
7. A. Avenoza, J. H. Busto, N. Canal and J. M. Peregrina, *J. Agric. Food Chem.* 2006, **54**, 4715–4720.
8. G. J. Martin, C. Guillou, M. L. Martin, M. T. Cabanis, Y. Tep and J. Aerny, *J. Agric. Food Chem.* 1988, **36**, 316–322.
9. J. T. W. E. Vogels, A. C. Tas, F.van der Berg, and J. A. van der Greef, *Chemom. Intell. Lab.* 1993, **21**, 249–258.
10. I. J. Kosir, and J. Kidric, *J. Agric. Food Chem.* 2001, **49**, 50–56.

SODIUM IONS IN MODEL CHEESES AT MOLECULAR AND MACROSCOPIC LEVELS

I. Andriot[1], L. Boisard[2], C .Vergoignan[2], C. Salles[2], E. Guichard[2]

[1]Lipids-Aromas Platform, Centre des Sciences du Goût et de l'Alimentation, UMR1324 INRA, UMR6265 CNRS, Université de Bourgogne, Agrosup Dijon, F-21000 Dijon.
[2]FLAVI Research Group, Centre des Sciences du Goût et de l'Alimentation, UMR1324 INRA, UMR6265 CNRS, Université de Bourgogne, Agrosup Dijon, F-21000 Dijon.

1. INTRODUCTION

The excessive consumption of sodium is one of the causes of nutritional-related health problems. The reduction of salt content without affecting technological and sensorial properties of foodstuffs is currently a challenge for the food industry.[1] There is a need to develop tools to quantify the "active" sodium ions in food products at molecular and macroscopic levels to better understand the in-mouth salt release. In this context, methods for the quantification of the bound fraction of sodium were developed.

2. MATERIALS AND METHODS

Model cheeses
Model cheeses with three different structures were made using a processed cheese technology[2] (65°C, 7 min 30, 2000 rpm) with three ratios of lipids/proteins (0.5, 0.75 and 1), and two salt concentrations (0 and 10 g.L^{-1} of NaCl). The water content was equal to 50% for all the cheeses studied.

^{23}Na NMR
At molecular level, the quantification of total sodium and bound sodium was performed by ^{23}Na NMR.[3, 4] A cylindrical sample ($\varnothing$ = 0.9 cm, length = 5 cm) was taken from a cheese sample and inserted in a 10 mm NMR tube. Another 5 mm NMR tube containing a solution of $Na_7Dy(PPp)_2$ in D_2O ([Na^+]= 0.4 M) was inserted as external reference.
^{23}Na single-quantum (SQ) and double-quantum-filtered (DQF) NMR experiments were recorded at 132.29 MHz on a Bruker Avance-500 spectrometer equipped with a 10 mm broad-band probe. All the NMR experiments were carried out at 25°C.

Partition coefficient
At macroscopic level, the partition coefficients of sodium ions in cheeses were determined according to the method developed by Lauverjat *et al.*[5] A piece of cheese (10 g) was put in 1 L of ultrapure water (MilliQ). After 24 hours of equilibrium at 25°C, an aliquot of water was taken. The quantification of sodium released in water was performed by a conductimetric detection, after ionic separation using HPLC (Dionex ICS-3000). Knowing

Table 1: *Quantification of total sodium and bound sodium, creation time and relaxation time by ^{23}Na NMR.*

Cheeses	Total sodium				Bound sodium		
	$[Na^+]$ (g.L^{-1})	Relaxation time (ms)		Ratio Area "bound Na"/ Area "Total Na"	Creation time (ms)	Relaxation time (ms)	
		T_{2S}^{SQ}	T_{2F}^{SQ}		τ_{opt}^{DQ}	T_{2S}^{DQ}	T_{2F}^{DQ}
Soft cheese without NaCl	6.91 ± 0.11	1.86 ± 0.01	5.65 ± 0.02	19.08 ± 0.58	1.60 ± 0.02	0.51 ± 0.01	9.81 ± 0.05
Soft cheese with NaCl	11.01 ± 0.06	1.73 ± 0.05	6.13 ± 0.03	18.6 ± 0.47	1.62 ± 0.00	0.51 ± 0.00	10.46 ± 0.00
Intermediate cheese without NaCl	7.21 ± 0.17	1.63 ± 0.29	4.42 ± 0.66	19.39 ± 0.05	1.19 ± 0.01	0.35 ± 0.01	8.96 ± 0.05
Intermediate cheese with NaCl	10.96 ± 0.16	1.43 ± 0.04	3.79 ± 0.08	20.37 ± 0.44	1.15 ± 0.44	0.34 ± 0.01	8.61 ± 0.19
Firm cheese without NaCl	7.26 ± 0.15	1.53 ± 0.05	3.92 ± 0.03	19.27 ± 0.29	1.14 ± 0.01	0.35 ± 0.00	8.10 ± 0.00
Firm cheese with NaCl	11.12 ± 0.21	1.43 ± 0.01	3.85 ± 0.01	19.30 ± 0.05	1.16 ± 0.00	0.35 ± 0.00	8.39 ± 0.03

the initial concentration of sodium in cheeses, the partition coefficients were calculated. Comparisons between the different obtained values were done using a Newman-Keuls test.

3. RESULTS AND DISCUSSION

Total sodium amount was quantified by ^{23}Na NMR with a good precision (Table 1). At molecular level, the relaxation time of total sodium (T_{2S}^{SQ}, T_{2F}^{SQ}) decreases when the firmness increases, thus the sodium mobility decreases. For an identical composition, when the sodium content increases, the sodium mobility decreases. Moreover, when the protein content increases, which induces an increase of firmness, the sodium ions are less mobile. Concerning the bound sodium, the ratio of "Area bound sodium/ Area total sodium" is approximately the same whatever the composition of cheeses.

The creation time of bound sodium (τ_{opt}^{DQ}) decreases when the cheeses firmness increases, thus the system is more organised around the bound sodium. The values of relaxation time for bound sodium (T_{2S}^{DQ}, T_{2F}^{DQ}) have the same behaviour as those of total sodium; i.e. the relaxation time of bound sodium decreases when the firmness increases, thus the bound sodium mobility decreases. For an identical composition, when the sodium content increases, the bound sodium mobility decreases.

Thus, the ^{23}Na relaxation time measurements and the DQF parameters highlight the impact of the composition (ratio of lipids/proteins) on the interactions between the sodium ions and the food matrices, confirming preliminary results obtained by *Gobet et al.*[4]

At macroscopic level, the partition coefficients (Table 2) are always lower than 1, which means that a part of sodium is retained by the cheese matrices. The partition coefficient of sodium decreases with the increase of cheese firmness. These results are in agreement with those obtained at molecular level.

Table 2: *Partition coefficient of sodium in cheese determined at 25°C.*

Cheeses	Ratio Lipids/Proteins	[Na+]$_{initial}$ (g.L^{-1})	Partition coefficient
Soft cheese without NaCl	1	7.2	0.79[a]
Soft cheese with NaCl	1	11.1	0.82[a]
Intermediate cheese without NaCl	0.75	7.2	0.76[b]
Intermediate cheese with NaCl	0.75	11.1	0.76[b]
Firm cheese without NaCl	0.5	7.2	0.69[c]
Firm cheese with NaCl	0.5	11.1	0.72[c]

[a,b,c]: significantly different at 0.05

4. CONCLUSION

This study allows to better understand the role of sodium in foodstuffs. It highlights that NMR is a powerful tool to quantify total sodium and bound sodium ions. Results on mobility at molecular level explain the differences in the macroscopic sodium release in water. A better understanding of sodium mobility in function of food composition and structure will help to better explain saltiness.

References

1.   D. Kilcast and F. Angus, eds., *Reducing salt in foods*, Woodhead Publishing Limited, Cambrigde, England (GBR), 2007.
2.   A. Tarrega, C. Yven, E. Sémon and C. Salles, *Int. Dairy J.*, 2008, **18**, 849.
3.   M. Mouaddab, L. Foucat, J.-P. Donnat, J.-P. Renou and J.-M. Bonny, *J. Magn. Reson.*, 2007, **189**, 151.
4.   M. Gobet, C. Rondeau-Mouro, S. Buchin, J.-L. Le Quéré, E. Guichard, L. Foucat and C. Moreau, *Magn. Reson. Chem.*, 2010, **48**, 297.
5.   C. Lauverjat, C. de Loubens, I. Déléris, I. C. Tréléa and I. Souchon, *J. Food Eng.*, 2009, **93**, 407.

THE IMPACT OF FREEZE-DRYING ON MICROSTRUCTURE AND HYDRATION PROPERTIES OF CARROT

Adrian Voda[1], Gerard van Dalen[1], Jaap Nijsse[1], Henk Van As[2], John van Duynhoven[1,2]

[1]Unilever Research and Development Vlaardingen, Olivier van Noortlaan, P.O. Box 114, 3130 AC Vlaardingen, The Netherlands
[2] Laboratory of Biophysics and Wageningen NMR Centre, Wageningen University, Dreijenlaan 3, 6703 HA Wageningen, The Netherlands

Corresponding author: Adrian Voda, (Olivier van Noortlaan, 3130 AC Vlaardingen, The Netherlands, *adrian.voda@unilever.com*, Phone: 31 10 4606708, Fax: 31 10 4606545)

1. INTRODUCTION

Lack of convenience in preparing healthy meals rich in fruits and vegetables is impeding consumers to reach their recommended daily intake. This issue can be addressed by providing the consumer with dried fruits and vegetables that can be rehydrated shortly before consumption. A major bottleneck for further growth in this area is that sensory properties are compromised due to drying technologies currently in use.

Most dehydrated fruits and vegetables are produced by hot air-drying. A disadvantage of this method is the substantial degradation in quality, including appearance (shrinkage, drying up, darkening), nutrients, flavor and the low rate of rehydration [2]. Higher quality products can be obtained using much more expensive freeze-drying methods. However, the resulting very porous structure leads to a loss in texture and increase in friability. Most developments in drying of fruits and vegetables have been engineering-driven. In typical engineering approaches, emerging new drying technologies have been optimized to obtain rehydration properties and texture that match as much as possible the original properties.

So far no systematic approach that considers the underlying microstructural events during the fresh-dried-rehydrated processing chain has been applied. A major barrier to embark on such an approach has been the lack of measurement adequate technologies that would enable decision making based on sound microstructural data. Hence we have deployed magnetic resonance spectroscopy and imaging to investigate microstructural impact of thermal pretreatments and freeze-drying. Since the microstructure upon drying determines the rehydration behavior we also assessed water distribution in this state.

We have employed a suite of complementary analytical techniques to investigate the structural impact upon freeze-drying as well as re-hydration behaviour and water distribution in the re-hydrated state. Different drying routes consisting in combination of drying methodology and various pretreatments have been exploited on series of carrot samples in order to explore the structure-damage relationship. At the μm-mm level, dried structures can be best assessed by X-ray microtomography (μCT) and scanning electron microscopy (SEM). μCT allows for high-resolution 3D visualisation and characterisation of dry and hydrated material [10]. One can probe the microstructure of samples non-invasively up to a few millimetres across with an axial and lateral resolution down to few micrometers. The contrast in μCT images is based on the difference in absorption of X-rays by the constituents of the sample (e. g. water and air). μCT allows observations under environmental conditions without sample disturbing preparations. SEM images reveal details of sample surface at sub-μm resolution as resulted from the interaction of an electron beam with atoms.

Since the microstructure upon drying determines the rehydration behaviour, time-domain NMR and MRI were used to assess the distribution and mobility of water in re-hydrated samples in a non-invasive manner. Compartment integrity and tissue permeability/tortuosity were studied by NMR diffusometry. Molecular diffusion (self diffusion) of water molecules in a re-hydrated plant material is restricted by semi-permeable surfaces which form pores with various sizes and shapes, connected by passages with different lengths and curvatures, all of these building up a complex microstructure. The diffusion coefficient of water in such systems is time-dependent and it has extensively been studied in the literature [7,8], and references therein. At short observation times the diffusion behaviour is determined by the surface-to-volume ratio of the pore space. At longer times, molecules probe the connectivity of the pore matrix but also the permeability of the pore walls[9].

2. METHODS

2.1. Carrot samples, pretreatments and drying

Figure 1 *Schematic representation of four distinct carrot tissue regions. The grey round mark indicates the sampling spot for the cylindrical cut samples employed in this work.*

Winter carrots were used in the current study, purchased in a local supermarket, and having sizes of 3 to 6 cm in diameter and a length of 20 to 30 cm. The carrot root consists of two distinct layers: a central stele, and peripheral cortex. The anatomy of the carrot root is greatly affected by the growth of the stele, hence four different regions can be distinguished as shown in Figure 1: three inner regions belonging to the stele (vascular tissue) and the outer region which is the cortex. Cylindrical samples of carrot cortical tissue with a diameter of 8 mm and a length of 10 mm were cut as it is shown in Figure 1.

Different thermal pretreatments were applied to the carrot samples before freeze-drying. The first pretreatment was blanching, which was performed for one minute in boiling water. Non blanched samples were also selected for the next step. Secondly, samples were frozen at four different temperatures: -28, -80, -150 and -196 °C.

2.2. Scanning electron microscopy (SEM)

A piece of dried carrot was cut into two halves in such a way that a cross-section was obtained. A very thin slice was cut off from surface with a razor blade to obtain a high quality cross section surface of the remaining piece of dry tissue. This surface was sputter coated with Platinum for better SEM imaging quality. Then the sample was inserted into a Jeol 6490 LA Scanning Electron Microscope and both the peripheral and central areas were imaged at standard magnification ranging from 10x to 1000x.

In the case of re-hydrated carrots, a small sample (max $2*2*2$ mm^3) was cut using a razor blade. The sample was placed on a rivet-type sample holder in a small drop of Tissue Tek cryo-adhesive and directly plunge-frozen in liquid ethane. The sample was stored in liquid nitrogen until further processing. In order to obtain a representative image of the tissue, the sample was cryoplaned in a cryo-ultramicrotome (Leica), first with a glass knife and the last sections with a diamond knife. After cryo-planing, the sample was stored in liquid nitrogen until SEM analysis. For SEM analysis, the sample was shortly sublimated (freeze-etched) at -90 °C in the cryo-preparation chamber (Oxford CT 1500HF) of the microscope (Jeol JSM6340F). After sublimation a very thin layer (few nanometers) of Au/Pd was sputtered onto the surface and the sample was transferred into the microscope for analysis. The sample was analysed at -125 °C, at 3 kV and a short working distance of about 6 mm, using the in-lens secondary electron detector (SEI). A more detailed description of this method can be found in [4].

2.3. X-ray microtomography (μCT)

The internal porous structures of the samples were visualised using a SkyScan 1072 desktop X-ray micro-tomography system (Belgium, http://www.skyscan.be) [5]. The X-rays were generated by a microfocus X-ray tube (10 μm focal spot size) with tungsten anode generating an energy ranging from 10 to 100 keV (peak at 80 kV). Power setting of 50kV and 100μA were used. The μCT produced two-dimensional images of projections of a sample [6]. The sample was imaged in a plastic cylindrical sample holder with an inner diameter of 11 mm. A stack of 973 flat cross sections was obtained after tomographic reconstruction of the images (1024 x 1024 pixels) acquired under different rotations over 180 degrees with a step size of 0.45 degrees. The acquisition time for one projection was 2.8 s resulting in a total acquisition and read-out time of about 25 min (413 projection images). The magnification of the projection was changed by changing the distance between the sample and the source. Magnifying the object allowed to increase the spatial resolution. A magnification factor of 23 was selected resulting in pixel sizes of 12 μm. The contrast in μCT images was based on the difference in absorption of the X-rays by the constituents of the sample (e.g. solids and air), which is influenced by the density and composition of the constituents.

2.4. Nuclear magnetic resonance (NMR)

2.4.1. Time-domain NMR relaxation experiments were carried out at room temperature (20° C) on a Bruker minispec operating at proton frequency of 20 MHz. Transverse relaxation times were measured using a Carr-Purcell-Meiboom-Gill (CPMG) pulse sequence with an echo-time (TE) of 0.2 ms and repetition time (TR) of 10 s. Transverse relaxation time, T_2, continuous distributions were calculated by means of ILT.

2.4.2. Magnetic resonance imaging (MRI) experiments were carried out at 7 T field strength (300 MHz proton operating frequency) using a Bruker Avance III spectrometer. The microimaging hardware employed consists of a 10 mm birdcage type resonator and a field gradient unit allowing for a maximum strength of 1.5 T/m. Re-hydrated carrot samples were placed in NMR glass tubes and experiments were performed at room temperature. Longitudinal and transversal morphological images (spin density) were recorded by means of 2D-RARE imaging [3] with a selected slice of 500 µm. In-plane field of view (FOV) was 12*12 mm^2. The complete *k*-space was covered in 128 phase encoding steps and 128 read points, resulting in a spatial resolution of 78 µm / pixel. The slice selection has been achieved with Gaussian shape radio frequency (r.f.) excitation/refocusing pulses of duration 1 ms/0.58 ms. TE was set at 3 ms and TR at 2 s with crusher gradients following each acquisition to destroy residual transverse magnetization from long relaxation components.

2.4.3. Molecular diffusion of water molecules in a re-hydrated plant material is restricted by semi-permeable surfaces that form pores with various sizes and shapes, connected by passages with different lengths and curvatures. Therefore, certain time-dependent diffusion behaviour exists in such systems. At short observation times (*t*), the diffusion coefficient *D(t)* is determined by the surface-to-volume ratio of the pore space. At long times, the molecules probe the connectivity of the pore matrix but also the permeability of the pore walls, and *D(t)* is determined by a tortuosity parameter weighted by the permeability of the re-hydrated cell wall material. The behaviour of the diffusion coefficient in time can be approximated by (Sen, 2004)

$$\frac{D(t)}{D_0} = 1 - \left(1 - \frac{1}{\Im}\right)\frac{\dfrac{4\sqrt{D_0 t}}{9\sqrt{\pi}}\dfrac{S}{V} + \left(1 - \dfrac{1}{\Im}\right)\dfrac{D_0 t}{D_0 \theta}}{\left(1 - \dfrac{1}{\Im}\right) + \dfrac{4\sqrt{D_0 t}}{9\sqrt{\pi}}\dfrac{S}{V} + \left(1 - \dfrac{1}{\Im}\right)\dfrac{D_0 t}{D_0 \theta}}, \qquad (1)$$

where S is pore surface area, V is pore volume, θ is a fitting parameter with dimensions of time. Equation (1) allows for the estimation of diffusion coefficient at 0 time, D_0, the surface-to-volume ratio of the pores and the tortuosity $\Im$, which is the asymptotic value of the equation in the long-time behaviour: $\dfrac{D(t \to \infty)}{D_0} \to \dfrac{1}{\Im}$.

The pulsed field gradient stimulated echo pulse sequence (PFG-StE) with bipolar gradients was employed to measure diffusion behaviour of water in re-hydrated carrots at different diffusion times. Experiments were carried out on the micro-imaging equipment described in the section *2.4.2*. The gradient duration was kept constant, 2 ms, while the gradient strength was varied between 0 and 1.5 T/m in 12 logarithmic distributed steps. The diffusion observation time spanned an interval between 20 and 1000 ms. Repetition time was set to 6 s and the signal was averaged in 16 scans.

3. RESULTS AND DISCUSSION

3.1. Effect of blanching, freezing temperature and freeze-drying on carrot properties

During freezing, ice crystals are formed from water present in the vacuoles and eventually surrounding tissue, thereby dehydrating the plant material around the crystals. Once freezing is complete, the tissue contains a number of ice crystals surrounded by dehydrated structures and sugars. Re-hydrated carrots previously frozen at different temperatures show an increase in texture with decreasing the freezing temperature before drying, as it is shown in Figure 2 a.

Blanching for a short time is loosening the cell membranes, increasing permeability and facilitating mass transport within the matrix. Applied before slow freezing, blanching affects the efficiency of ice crystal formation, therefore smaller pores and fewer cracks are expected in these samples and blanching leads to an increase of texture (Figure 2 a). In the case of fast freezing, blanching leads to overall cellular matrix acceptance of being frozen, meaning that homogenization of sugar concentration offers an overall cryo protection while mechanical stress is minimized due to walls loosening. In this case texture was observed to decrease as compared to non-blanched carrots.

3.2. Microstructural features of FD carrots by SEM and μCT

Information on the state of the dry microstructure and the integrity of the cellular tissue can be collected from SEM images shown in Figure 2 b. The sublimation of the ice crystals grown within the carrot leaves a dried matrix representing a fingerprint of the crystals size and shape. Samples frozen at lower temperature show smaller pores as the ice crystals are expected to grow smaller under faster cooling conditions. However, during freezing, the crystal growth ruptures, pushes and compresses cells together. This damage is more pronounced in slow frozen tissue where the ice crystals grow bigger. Moreover, the image of the fast frozen sample clearly reveals a certain cellular architecture that is very similar to the native cellular tissue. The microstructures of complete dried carrot pieces were imaged by μCT (Figure 2 c). The overall difference in pore sizes can be observed. However, quantitative image analysis is needed in this case for porosity and connectivity estimation. The direction of the ice crystal formation can be clearly seen. Moreover, sampling the carrot pieces in cases when two (or more) distinct tissue zones are subject to thermal treatment and drying result in the appearance of randomly oriented mechanical stress and causes additional microstructure damage.

3.3. Structure and morphology of re-hydrated carrots by TD-NMR and MRI

MRI and SEM were employed to visualize water distribution in re-hydrated carrot samples. At sub-μm level, SEM images (not shown here) reveal the hydration/swelling status of individual cells and cell walls. The signal intensity in the MRI images is directly proportional with the water content. The carrot sample that was frozen at -196°C reflects a native structure pattern after re-hydration (Figure 2 d).

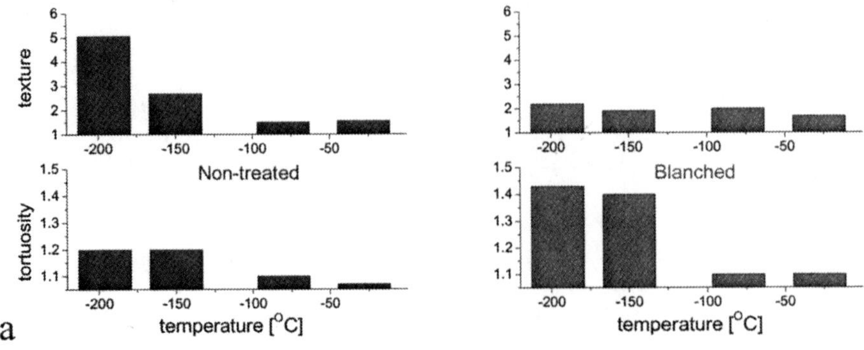

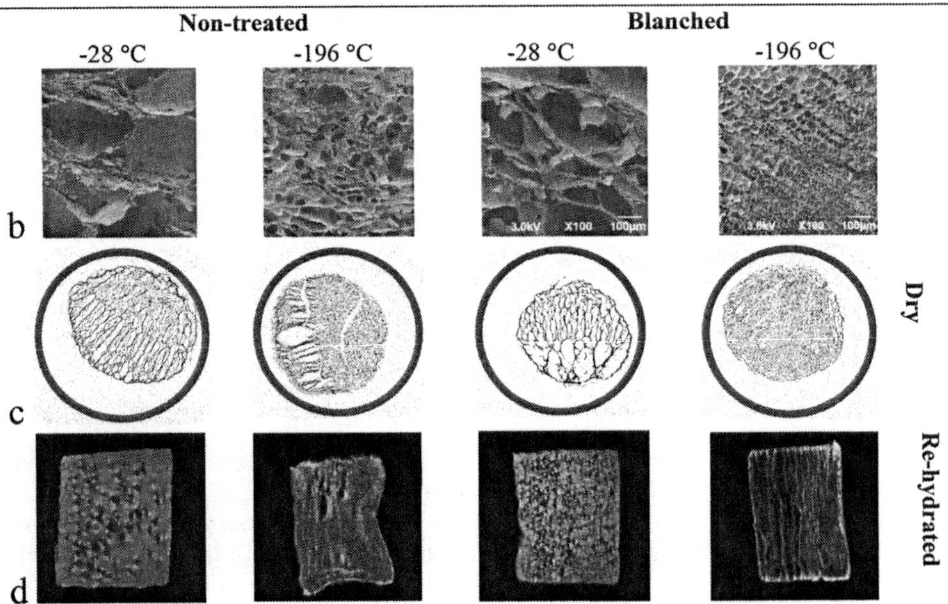

Figure 2 *Micro- and macroscopic features of pre-treated freeze-dried and re-hydrated cortical tissue of carrots. a: texture (compression tests) and tortuosity (from water self diffusion by NMR) in re-hydrated state; b, c: SEM and μCT images of dry carrot structures; d: MRI of re-hydrated carrots.*

Additional information about the microstructure of carrots can be obtained from NMR relaxation and diffusion measurements. The NMR transverse relaxation experiments on fresh carrots reveal the main water populations present in the vacuoles and in the cell walls. Longer relaxing components have been identified in re-hydrated samples frozen at -28°C that relate to the water filling cavities larger then the original cell size. During blanching, the vacuoles and organelles present in the cell will either show a collapse or they might maintain original shape but not the functionality. Loosening of cell walls and increasing membranes permeability is an immediate consequence of heat treatment [1]. Therefore the impact of freezing on a raw and blanched material is different. Having in mind that blanching occurs only for one minute, probably the sugar concentration in the whole carrot sample is not significantly reduced but the ingress of water occurs from outside into the empty (air) spaces within the carrot tissue, or/and water release from the vacuoles into the extracellular space. Hence, a process of solutes homogenization takes place within the carrot tissue. Though the cellular architecture can look very similar, a blanched tissue can be regarded more as a sponge immersed in water.

To get more insight into the impact of blanching and freeze-drying on the microstructure of carrot pieces, relaxation behaviour was analyzed in carrot samples upon 15 minutes of re-hydration in hot water (>90°C). The distribution of relaxation times in re-hydrated carrots is shown in Figure 3. We formulated a hypothesis to assign different peaks and explain differences between samples. Long relaxing components, above 400 ms, have been identified in samples frozen at -28°C that relate to the water filling cavities larger then the original cell size. This is consistent with the ice crystal growth effect on the carrot structure. The group of relaxation peaks between 200 and 400 ms corresponds to the water population filling pores similar to the cell size, as the vacuolar water species of the fresh carrot can also be identified in this region of the spectrum (data not shown). The faster relaxing peak of the blanched sample frozen at -196°C is a unique feature among re-hydrated samples. This can be due to water present in the swollen cell walls, as this sample has shown very similar cellular architecture with the fresh tissue (based on SEM images). However, this remains to be further investigated by two-dimensional relaxation/diffusion experiments.

The knowledge on the microstructure of re-hydrated carrots can be complemented by analyzing self-diffusion behaviour of water over a variable time scale, which is translatable into distances traveled by water molecules within the porous media. In fresh plant material semipermeable membranes exist and physical quantities like compartment sizes and permeability can be determined via the numerical plant cell model [11]. Processing like blanching and drying result in a significant increase in membrane permeability, which makes the re-hydrated system to behave more like a sponge characterized by a certain tortuosity. Self diffusion coefficients of water in rehydrated carrot pieces were measured along axial and transverse directions over a diffusion time range up to 1 s. Anisotropic diffusion behaviour was observed in all samples (data not shown). Samples frozen at low temperature show more anisotropy, which again confirms structure preservation. By fitting the time dependent diffusion constant curves with equation (1), tortuosities were obtained and are shown in Figure 2 a. The tortuosity of slow frozen samples at -28°C and -50°C is very close to unity, which means that water molecules can easily travel through the porous structure with rather no obstructions. This is a clear indication that no authentic structure features of a native carrot exist in these samples, and so, any processing step that facilitates the ice crystal growth will have an irreversible damage. Higher tortuosity and texture (Figure 2 a) were determined in fast frozen carrots. The texture of blanched tissue is inferior, as the strength of the cell walls has decreased. On the opposite, the tortuosity is higher. This suggests that blanching reduces the structure damage during fast freezing. In a raw carrot the distribution of water, sugars and extracellular air spaces is heterogeneous. Blanching may homogenize the solutes concentration and release the air, therefore mechanical stress is well diminished. Additionally, fast freezing of such a homogenized network induces less structural changes than freezing a fresh plant material. Though tissue architecture and water pathways may be better preserved by blanching, the texture of the whole piece is reduced.

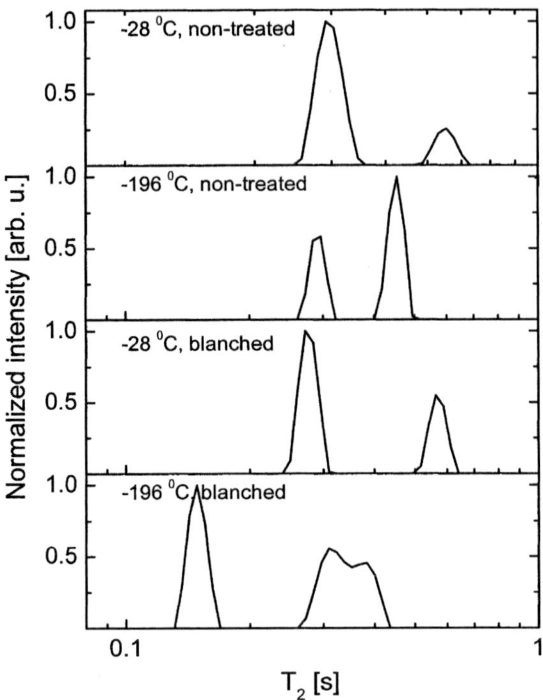

Figure 3 *Transverse relaxation times distribution in re-hydrated carrot showing the water populations that reveal the presence of cellular-size pores and bigger cavities.*

4. CONCLUSIONS

An effective combination of imaging techniques was demonstrated to generate a comprehensive multi-scale definition of fruits and vegetables microstructure. The complexity of phenomena during thermal pretreatments and freeze-drying of carrots was resolved by complementary information generated by these techniques. We have used this approach to analyze the impact of thermal pretreatments and drying on the microstructure of cortical tissue of winter carrots. It was concluded that blanching determines the formation of larger ice crystals in the case of slow freezing and a more homogeneous freezing with less structural damage in the case of fast freezing. Slow freezing allows ice crystals to grow outside cells, causing damage by cell collapse and rapture. Fast freezing determines ice crystals to grow inside cells with very little cell separation and much less damage. Freeze-drying following the above pretreatments does not further change the structure of the frozen tissue.

References

1. Gonzalez, M. E., Barrett, D. M., McCarthy, M. J., Vergeldt, F. J., Gerkema, E., Matser, A. M., and Van As, H., 1H-NMR Study of the Impact of High Pressure and Thermal Processing on Cell Membrane Integrity of Onions, *Journal of Food Science*, 2010, **75**, 7, E417-E425,.

2. Jayaraman, K. S. and Das Gupta, D. K., *Drying of fruits and vegetables*, 2007.

3. Matt A.Bernstein and Kevin F.King, Xiaohong Joe Zhou, *Handbook of MRI Pulse Sequences*, 2004.

4. Nijsse, J. and van Aelst, A. C., Cryo-planing for cryo-scanning electron microscopy, *Scanning*, 1999, **21**, 6, 372-378.

5. Sasov, A. and Van Dyck, D., Desktop X-ray microscopy and microtomography, *Journal of Microscopy-Oxford*, 1998, **191**, 151-158.

6. Sasov, A. Y., Microtomography .1. Methods and Equipment, *Journal of Microscopy-Oxford*, 1987, **147**, 169-178.

7. Sen, P. N., Time-dependent diffusion coefficient as a probe of geometry, *Concepts in Magnetic Resonance Part A*, 23A, 2004, **1**, 1-21.

8. Sibgatullin, T. A., Vergeldt, F. J., Gerkema, E., and Van As, H., Quantitative permeability imaging of plant tissues, *European Biophysics Journal with Biophysics Letters*, 2010, **39**, 4, 699-710.

9. Van As, H., Intact plant MRI for the study of cell water relations, membrane permeability, cell-to-cell and long distance water transport, *Journal of Experimental Botany*, 2007, **58**, 4, 743-756.

10. van Dalen, G. Notenboom P. van Vliet L. J. Voortman L. & Esveld E., 3-D Imaging, Analysis And Modelling Of Porous Cereal Products Using X-Ray Microtomography, *Image Analysis Stereology*, 2007, **26**, 169-177.

11. van der Weerd, L., Melnikov, S. M., Vergeldt, F. J., Novikov, E. G., and Van As, H., Modelling of self-diffusion and relaxation time NMR in multicompartment systems with cylindrical geometry, *Journal of Magnetic Resonance*, 2002, **156**, 2, 213-221.

New Developments in NMR

MEDIUM RESOLUTION NMR AT 20 MHz: POSSIBILITIES AND CHALLENGES

M. Cudaj[1,2], T. Hofe[1], M. Wilhelm[2], M. A. Vargas[3], G. Guthausen[3]

[1]PSS, In der Dalheimer Wiese 5, 55120 Mainz,
[2]Institute for Chemical Technology and Polymer Chemistry, KIT, 76131 Karlsruhe,
[3]SRG 10-2, Institute for Mechanical Engineering and Mechanics, KIT, 76131 Karlsruhe,
Germany

1 INTRODUCTION

Detailed studies of compositions, for example in food and oil industries[1-4], are of great importance to guarantee a high and constant quality and well defined properties of a product. Many applications still need expensive analytical techniques like high resolution NMR (Nuclear Magnetic Resonance) to determine each component's concentration like the marginal amounts of acids, colorants or polyphenols[5] in wine analysis[6] or small amounts of special fatty acids in virgin olive oils[7]. Moreover high field NMR delivers valuable information for characterizing the botanical origin or for investigating the age of a natural product like vinegar[8,9].

Up to now, low field NMR instruments in the range of 1 – 60 MHz are almost exclusively applied in relaxometry studies[10,11], which proved to be a powerful method, for example, to characterize the movement and redistribution of water within soils on the basis of T_2 relaxation times of the water proton signal. However, there have been very few investigations on the ability of low field instruments delivering spectral information, concerning chemical shifts and J-coupling of the studied sample.

In contrast to relaxometry methods for determination of water and oil content[12-17], prerequisite for frequency resolved NMR proton spectra is a sufficient spectral resolution of the NMR spectrometer in the range of 0.2 ppm full width half maximum (FWHM)[18] and frequency or B_0 field stability. The general feasibility of NMR spectroscopy with adequate resolution at low magnetic fields have already been shown by different groups [19-22]. The technique is named medium resolution NMR (MR-NMR) in the following.

Apart from composition determination via MR-NMR, processes like fermentation, ripening or encapsulation of active agents are of interest. Reactant concentrations can principally be followed as a function of time allowing the study of reaction kinetics.

Moreover, a question which has to be addressed is the way how raw data are processed; the possibilities include integration or the statistical chemometric approach. Whether or not the limited spectral resolution and sensitivity at low magnetic fields can be "compensated" more effectively by "classical" data processing (integration and subsequent linear regression) or multivariate data analysis (chemometrics) is studied on two models. The ability of chemometrics as a tool for analysis of high field NMR spectra of mixtures has already been shown[23].

2 INSTRUMENTATION

Apart from a modified Bruker the minispec mq20 for on-line spectroscopy, the instrument used for these measurements was a medium resolution (MR-NMR) prototype from Bruker (Bruker BioSpin GmbH, Rheinstetten, Germany), which works at a proton (^{1}H) Larmor frequency of about 20.1 MHz and is based on a permanent magnet[18]. The magnet temperature is stabilized at 40°C within ± 0.001 °C. With passive, mechanical shimming the homogeneity of the magnet system is in the order of 1 ppm for a cylindrical sample volume of about ID 4 mm x 12 mm. Furthermore, the spectrometer features an electronic shim system with 12 channels with adjustable currents in the range of ± 100 mA. With this planar shim system a field homogeneity of around 0.1 ppm (FWHM), sufficient for MR-NMR, can be achieved.

The MR-NMR prototype was equipped with a home built probe with a quality factor $Q = 140$ for sample tubes with an outer diameter of 5 mm and with a 13 mm sensitive region along y. The optimum $\pi/2$ flip angle was obtained at 3 µs (transmitter power of about 5 W). Data acquisition with application of FIR (finite impulse response) digital filters is provided, involving an enormous data reduction factor due to on-the-fly oversampling. This leads to a significant *S/N* advantage once the improvement in homogeneity leads to longer FID (free induction decay) decay times in the order of several hundreds of ms.

3 IDENTIFICATION AND QUANTIFICATION

Conventional high field NMR spectroscopy is frequently used for identification and quantification of (unknown) substances. However, its spectral resolution is far better than in MR-NMR. In the following, some dedicated examples show the use of MR-NMR for identification and quantification.

3.1 Fatty acids composition

A variety of different oils, like sunflower or olive oil, were analyzed by ^{1}H-NMR spectroscopy with a 500 MHz NMR system[7]. NMR spectroscopy provided a possible alternative to conventional chromatographic methods for determining the chemical composition of oils.

With the 20 MHz spectrometer, spectra of rape oil and sunflower oil were acquired, both fats containing large amounts of unsaturated fatty acids. In contrast, a spectrum of Palmin® was recorded; a fat that contains "only" saturated fatty acids like palmitin acid due to hydrogenation. Figure 1 shows spectra of the three different oils respectively solid fat (molten at 40 °C, which is the sample temperature due to the tempering of the magnet at 40 °C). All spectra were generated by recording 128 scans and folding the practically noise free FID with Gauss-Lorentzian functions. The recycle delay time amounted to 1 s while the acquisition time was chosen as 0.67 s.

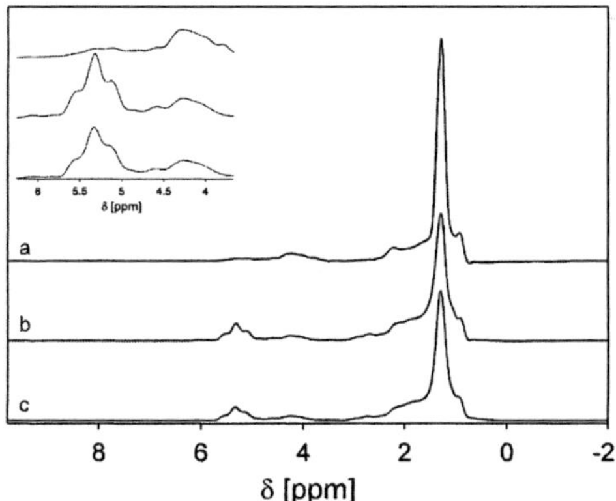

Figure 1: *¹H-spectra of Palmin® (a), sunflower oil (b) and rape oil (c). The spectra show relatively unspecific lines in the aliphatic spectral region (1-2 ppm). The significant peaks of carbon double bonds, which appear at 5.3 pm in spectra b and c, are a factor of 5 less in spectrum a.*

In the aliphatic spectral region (1-2 ppm) the different fat compositions do not differ significantly. In contrast the olefinic region is characteristic for each sample. Obviously, the significant peak of carbon double bonds, which appears around 5.3 ppm in the spectra of rape oil and sunflower oil, is clearly diminished in the spectrum of Palmin®. The spectral fraction of unsaturated fatty acids is estimated to 8.1 % (rape oil), 9.8 % (sun flower oil) and 2.2 % (Palmin®), which is in good agreement with the compositional expectation.

3.2 Pyrolysis oils from renewable natural resources

A field of large scientific interest is currently the energy generation from plants, especially from straw of food plants and woods. A major problem is the high and varying water content during biomass gasification which complicates the measurement and control process. A fast and reliable technique is needed which is additionally robust against environmental influences.

It has to be proven, that MR-NMR is a well-suited technique, by its ability to measure the water content of the pyrolysis oils. Figure 2 shows the MR-NMR spectra of oils with different botanical origin and pyrolysis processes. Apart from the water peak around 4.6 ppm, also the chemical composition of the oils varies significantly, which can be seen most obviously in the aliphatic region. Depending on the viscosity of the oils, the spectral signatures vary in their line width so that a chemical composition determination is not easily possible. However, the water line is relatively pronounced. Its integration was correlated with the water content determined by wet chemical procedures. A satisfying agreement was obtained with a coefficient of determination $R^2 = 0.99$ (0.97, when considering the outlier).

The spectra were measured averaging 4 scans with a repetition time of 10 s, a dwell time of 163.5 µs and 8k acquisition data points. During data processing, zero-filling by a factor of two was applied.

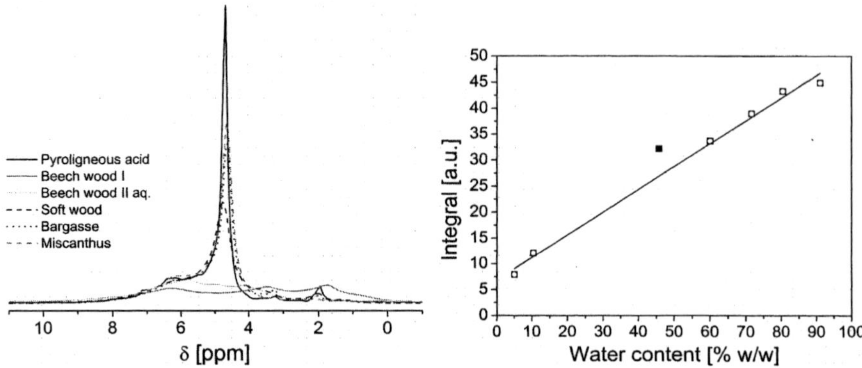

Figure 2: *The water content of various pyrolysis oils can be clearly identified in the ¹H-NMR spectra. In spite of their different chemical composition which is also observable over the spectral range, a correlation between the integral of the water peak around 4.6 ppm and alternatively determined water content can be established. The coefficient of determination amounts to 0.99.*

3.3 Concentration Determination in Binary Compounds

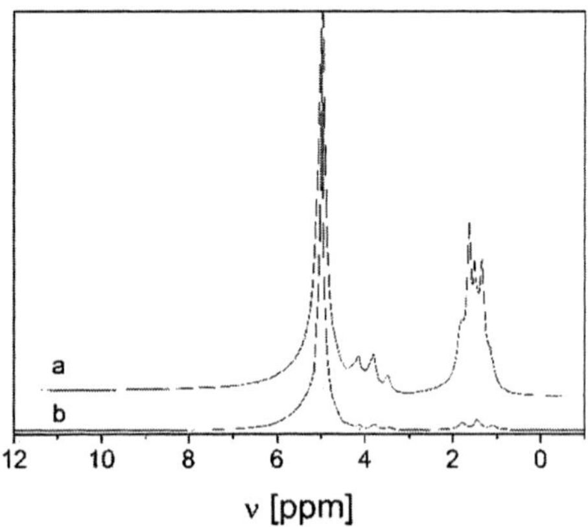

Figure 3: *Spectra of ternary ethanol water isopropanol mixture (30 vol.% ethanol, 30 vol.% isopropanol, a) and of binary ethanol water mixture (14 vol.% ethanol, b). The spectra are highly overlapping due to B_0 inhomogeneity and the relatively large J-coupling compared to chemical shift. Clearly distinguishable are the CH₃-lines at 1-2 ppm, the CH₂-groups show up between 3 and 4 ppm. The dominant peak in the spectra is due to water.*

Currently, wine quality analysis is often performed by wet chemical procedures (including e. g. Cr^{6+}) or infrared methods. Comparing the efforts of wet chemical methods with those required by low field NMR spectroscopy, MR-NMR could be a valuable tool for reducing efforts in ethanol determination. Compared with IR, the higher selectivity of NMR could be of advantage. However, accuracy and reproducibility have to be proven.

Figure 3 shows the spectra of a ternary ethanol water isopropanol mixture (a, 30 vol.% ethanol, 30 vol.% isopropanol) and a binary ethanol water mixture (b, 14 vol.% ethanol) at

20 MHz. The spectra were recorded with 8 scans for sufficient signal to noise ratio, 8 k data points and a zero filling factor of 4 in order to assure an adequate resolution. The FID acquisition time amounts to about 1.34 s, while the recycle delay time was 25 s guaranteeing complete recovery of the magnetization (5 T_1). To avoid further peak broadening due to magnetic field drifts and fluctuations, the spectra were frequency-corrected during data processing.

Compared to commonly known high field spectra of these substances, the lower relation of chemical shift to *J*-coupling at 20 MHz leads to different shapes and often considerable broadening of single, chemical shift separated peaks. Both the larger B_0 inhomogeneity and the larger *J*-coupling (in ppm) lead to highly overlapping signals. The question to be answered is to which extent larger peak widths and insufficient baseline separation of MR-NMR spectra affect the analysis of binary and ternary compounds in „classical" and multivariate data processing. A second question is which data processing method results in reproducible and accurate concentration determination. Apart from classical integration of specific peaks, multivariate data analysis is applied in form of PLS (partial least squares) regression (e.g., introduction into PLS-R[24]).

An accurate determination of concentrations from NMR spectra requires absolutely reproducible baselines, which are preferentially completely flat. Therefore, a section wise base line correction was performed. Integration limits were chosen in a way that the influence of neighbouring lines of the other moieties was minimal; on the other hand the sections should be still meaningful for the concentration determination.

PLS regression was accomplished using the TOMCAT tool[25]. It is based on Matlab® and is an open source program[25]. Cross validation was performed leaving out two samples of 36 calibration samples (+ 4 test set samples) with a maximum of 1000 iterations resulting in the calibration model in figure 4 (left). The large number of iterations was chosen because of increased numerical stability of the model. The intercept as well as the slope of the resulting linear regression equation $y = 0.99997x + 0.0006$ indicate a good correlation between "balance-determined" (x) and "measured" values (y) for ethanol concentration. Additionally, figure 4 shows the results for three test samples (right, x) in the calibration model as well as a residual plot. The optimum for the calibration was found at a rank of 10. When comparing the values of additional 200 MHz data with MR-NMR data (Table 1), one can assert that the correlation coefficient R as well as the model characterizing errors *RMS* (root mean square error), *RMSECV* (root mean square error of cross validation) and *RMSEP* (root mean square value of prediction) are in the same range for high and low field measurements. For the quality of measurement and model, the *RMSEP*, indicating an average absolute prediction error, is the most important value, which is in the range of 0.1 to 0.2%. It has to be compared with the required accuracy which is < 0.25 vol.% for German wines at the moment. This result implies the possibility of determining ethanol contents, e. g., in wine analysis by MR-NMR spectroscopy with the help of multivariate data processing.

The conventional quantification method in NMR spectroscopy is the integration of specific peaks. The assumption is that the integral of a well separated line can be directly correlated to the number of spins of a specific moiety, in this case of ethanol or water. Spectra were analyzed so that the integral of the CH_3-peak, which corresponds to half of the ^1H spins of ethanol, was taken as a measure for ethanol concentration. The volume concentration was calculated by consideration of molar mass and mass density of water and ethanol, without regarding the fact of volume contraction due to negative partial volumes when mixing water and ethanol. Baseline corrections were performed aiming for an accurate deter-mination of the methyl group integral and of the entire integral due to higher precision in the choice of integration limits. As in multivariate data analysis, the same three of the 40

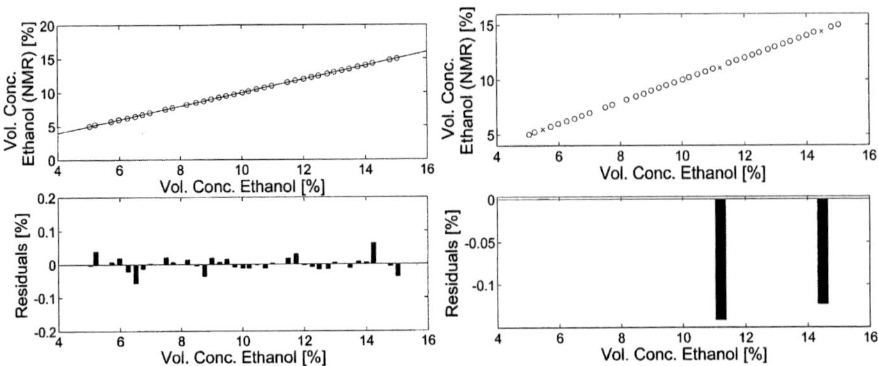

Figure 4: *Calibration (left, o) and test set validation (right, x) for ethanol quantification, obtained by applying a PLS regression to the* ^{1}*H-MR-NMR data. The ethanol determination is therefore regarded as being quantitative with rather small errors, which are indicated by the residuals for each sample in the calibration and the test sets, respectively. The coefficient of determination in the correlation plot amounts to* $R^2 = 0.9999, RMSEP = 0.11$ *vol.%.*

Table 1: *Regression results of 20 MHz and 200 MHz data, processed by applying PLS-R and integration.*

	20 MHz, PLS	200 MHz, PLS	20 MHz, integration	200 MHz, integration
Linear regression	$y=0.99995x+0.0006$ (0.997x+0.029)	$y=0.998x+0.019$	$y=0.661x+1.546$	$y=0.825x+2.260$
R	0.99997 (0.9986)	0.999	0.947	0.955
RMS	0.021 (0.157)	0.128	0.669	0.767
$RMSECV$	0.041 (0.165)	0.207	-	-
$RMSEP$	0.108* (0.198)	0.214	1.150	1.357

samples were left out from the linear regression model as they were needed as test samples and calculation of *RMSEP*. Neither linear regression nor mean prediction error could reach the quality of their chemometric analogues (Table 1). The "classical" data procedure is clearly less accurate and reliable compared to multivariate PLS-R analysis and could not reach required accuracy limits (*RMSEP* < 0.25 vol.%). Baseline drifts, imperfect shim currents, insufficient automated phase corrections and others led to this impreciseness, which finally favours chemometric methods for the investigation of binary compounds in the 5-15% range of absolute ethanol content.

Concluding, the spectroscopic measurement of binary mixtures of ethanol and water by MR-NMR spectroscopy, combined with chemometric data processing, shows its potential in quantitative determination of relatively high concentrated substances. Compared with 200 MHz, the accuracy is in the same order of magnitude for the high concentration samples.

4 ON-LINE PROCESS CONTROL

In process analytics material changes are of great interest. Processes like ripening, encapsulation of active agents and chemical reactions should preferentially be followed on-line. Analytical techniques like infrared spectroscopy and ultrasound measurements partially fulfil this need for time-resolved analyst concentration monitoring. An alternative could be MR-NMR with its inherent high chemical selectivity. First examples are described in literature[17, 20-22].

In this work the modified minispec mq20[19] was used to monitor the reaction progress in an emulsion polymerization[26]. This reaction type is of special interest in food processing science because of the above mentioned encapsulation possibility of active agents in droplets. Exemplarily, polymerization process of butyl-acrylate was monitored. From the NMR point of view the essential reaction is a carbon reduction from a sp^2- to a sp^3-hybridized state. Additionally, the sensitivity of transverse relaxation towards molecular mobility has to be considered.

In figure 5 the spectrum at reaction start is shown. As butyl-acrylate has sp^2-hybridized carbon atoms, a line around 6.3 ppm can be observed. The decrease of its intensity is a direct measure for the reaction progress. Additionally, residual H_2O and aliphatic components can be clearly distinguished. The line at the right is due to the reference in the probe, which is used as an independent measure for frequency correction and reproducibility check. This is necessary because of magnetic field drifts on the short- and long term scale. The reaction was performed in D_2O because of the dominant H_2O peak.

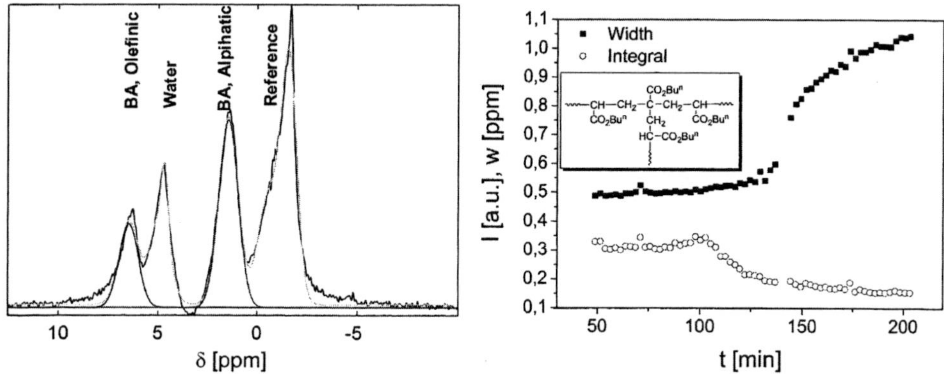

Figure 5: *Typical spectrum of butyl-acrylate in D_2O (left) as measured with a modified mq20. The olefinic line at 6.3 ppm decreases in amplitude whereas in the aliphatic region an increase of line width and amplitude can be observed. The line around 4.6 ppm is mainly due to residual H_2O, the peak at -2.0 ppm comes from reference sample in the probe. On the right hand side the integral of the olefinic region and the line width in the aliphatic region are shown as a function of reaction time.*

All spectra were fitted in a MATLAB® routine resulting in graphs for intensities and widths[26]. The significant parameters are the intensity of the olefinic peak and the line width in the aliphatic region. These two parameters can be described by kinetic models, revealing typical rate constants for the emulsion polymerization[26]. Please note the time shift between the start of integral decrease and width increase, which is caused by different physical phenomena reflected by the two NMR spectral parameters.

The reaction can also be followed when performed in H_2O. However, due to the limited spectral resolution of the current instrumentation, the error bars of the olefinic intensity are rather large. Nevertheless, the line width in the aliphatic region can be used for numerical analysis for determination of the kinetic coefficients.

5 CONCLUSIONS

MR ^{1}H-NMR spectra (20 MHz, resolution < 0.2 ppm) can be a real alternative to conventional analytical tools, even more in combination with multivariate data analysis. When comparing the MR-NMR spectra with those measured at a 10 times higher field strength on samples with sufficiently high concentrations, it could be shown that data processing by multivariate analysis tools is an essential element that can improve prediction ability, accuracy and reproducibility of an analyzed system. The limiting factors are for highly concentrated samples not the signal to noise ratio nor the spectral resolution, but artefacts like phase and baseline errors as well as spurious signals.

In addition to concentration and identification studies, MR-NMR spectroscopy can also be used as a chemically sensitive method in reaction studies and control. It provides detailed insight into the reaction kinetics.

Acknowledgements

M. C. and G. G. would like to thank Bruker BioSpin GmbH (especially A. Kamlowski and D. C. Maier) for allocation of the MR-NMR prototype spectrometer as well as Bruker BioSpin development team for valuable discussions and speedy help. The 'Shared Research Group 10-2' received financial support by the 'Concept for the future' of Karlsruhe Institute of Technology within the framework of the German Excellence Initiative. Additional support was given by FhG ICT and DFG, which are also highly appreciated. Financial support by the Investitions- und Strukturbank Rhineland-Palatinate (ISB) GmbH" is gratefully acknowledged. M. V. acknowledges the financial support by the DAAD.

References

Please consider also the references in the cited publications.

1 I. F. Duarte, A. N. Barros, C. U. Almeida, M. Spraul, A. M. Gil, *J. Agric. Food Chem.*, 2004, **52**, 1031.
2 M. R. Monteiro, A. R. P. Ambrozin, L. M. Lião, E. F. Boffo, E. R. Pereira-Filho, A. G. Ferreira, *J. Am. Oil. Chem. Soc.*, 2009, **86**, 581.
3 M. R. Monteiro, A. R. P. Ambrozina, M. da Silva Santos, E. F. Boffo , E. R. Pereira-Filho, L. M. Lião, A. G. Ferreira, *Talanta*, 2009, **78**, 660.
4 M. R. Monteiro, A. R. P. Ambrozin, L. M. Liao, E. F. Boffo, L. A. Tavares, M. M. C. Ferreira, A. G. Ferreira, *Energy & Fuels*, 2009, **23**, 272.
5 F. H. Larsen, F. van den Berg, S. B. Engelsen, *J. Chemometrics*, 2006, **20**, 198.
6 G. E. Pereira, J.-P. Gaudillere, C. van Leeuwen, G. Hilbert, O. Lavialle, M. Maucourt, C. Deborde, A. Moing, D. Rolin, *J. Agric. Food Chem.*, 2005, **53**, 6382.
7 C. Fauhl, F. Reniero, C. Guillou, *Magn. Reson. Chem.*, 2000, **38**, 436.

8 R. Consonni, L.R. Cagliani, F. Benevelli, M. Spraul, E. Humpfer, M. Stocchero, *Analytica Chimica Acta*, 2008, **611**, 31.

9 G. J. Martin, C. Guillou, M. L. Martin, M. T. Cabanis, Y. Tep and J. Aerny, *J. Agric. Food Chem.*, 1988, **36**, 316.

10 T. R. Todoruk, M. Litvina, A. Kantzas, C. H. Langford, *Environ. Sci. Technol.*, 2003, **37**, 2878.

11 C. A. Toussaint, F. Médale, A. Davenel, B. Fauconneau, P. Haffray, S. Akoka, *J. Sci. Food Agric.*, 2002, **82**, 173.

12 H. T. Pedersen, L. Munck, S. B. Engelsen, *J. Am. Oil. Chem. Soc.*, 2000, **77**, 1069.

13 J. Brùndum, L. Munck, P. Henckel, A. Karlsson, E. Tornberg, S. B. Engelsen, *Meat Science*, 2000, **55**, 177.

14 S. M. Jepsen, H. T. Pedersen, S. B Engelsen, *J. Sci. Food Agric.*, 1999, **79**, 1793.

15 A. Guthausen, G. Guthausen, H. Todt, W. Burk, A. Kamlowski, D. Schmalbein, *J. Am. Oil. Chem. Soc.*, 2004, **81**, 727.

16 G. Guthausen, J. König, A. Kamlowski, *Bruker Spin Report*, 2004, **154**, 41.

17 A. Nordon, P.J. Gemperline, C.A. McGill, and D. Littlejohn, 2001, **73**, 4286.

18 M. Cudaj, G. Guthausen, A. Kamlowski, D. Maier, T. Hofe, M. Wilhelm, to appear in *Nachrichten aus der Chemie*, November 2010.

19 G. Guthausen, A. von Garnier, R. Reimert, *Applied Spectroscopy*, 2009, **63**, 1121.

20 A. Nordon, A. Diez-Lazaro, C. W. L. Wong, C. A. McGill, D. Littlejohn, M. Weerasinghe, D. A. Mamman, M. L. Hitchman, J. Wilkie, *Analyst*, 2008, **133**, 339.

21 T. W. Skloss, A. J. Kim, J. F. Haw, *Analytical Chemistry*, 1994, **66**, 536.

22 http://www.qualion-nmr-com

23 H. Winning, F. H. Larsen, R. Bro, S. B. Engelsen, *J. Magn. Reson.*, 2008, **190**, 26.

24 P. Geladi and B. R. Kowalski, *Analytica Chimica Acta*, 1986, **185,** 1.

25 M. Daszykowski, S. Serneels, K. Kaczmarek, P. van Espen, C. Croux, and B. Walczak, *Chemom. Intell. Lab. Syst.*, 2007, **85**, 269.

26 M. A. Vargas, M. Cudaj, K. Hailu, K. Sachsenheimer, and G. Guthausen, *Macromolecules*, 2010, **43**, 5561.

RAPID AND VALIDATED NMR QUANTIFICATION APPROACHES FOR COMPLEX METABOLITE MIXTURES

B. Braganti[1], S. Peters[1,3], D.M. Jacobs[1,3], T. Eymond[1], M. Klinkenberg[1,3], J. van Duynhoven[1,2,3]

[1]   Unilever R&D, Vlaardingen, The Netherlands
[2]   Wageningen University, Wageningen, The Netherlands
[3]   Netherlands Metabolomics Centre, Leiden, The Netherlands

1 INTRODUCTION

NMR has the unique advantage that it can rapidly detect and quantify low-molecular weight molecules in complex mixtures with minimal sample preparation procedures, typically addition of an internal standard and lock solvent. Recent advances in NMR hardware have increased performance with respect to sensitivity and long-term stability. This opens up applications of NMR for quantification in complex mixtures at concentrations and sample throughput that was hitherto not considered feasible. Hence, we carried out a systematic validation of rapid and automated NMR quantification methods for samples taken from *in-vitro* models for bioconversions of food ingredients by colonic microbiota[1,2].

2 METHODS

Measurements were performed on a 600 MHz Bruker Avance III spectrometer, equipped with a 5 mm cryoprobe. Throughout the study small-volume (200 µl) 3 mm SampleJet tubes were used. Quantification relied on Internal (IS) and External[3] (ES) Standards methods. Accuracy was determined with respect to weighed-in concentrations, and levels as determined by GCMS[4]. Accuracy was expressed as proportional bias and offset as determined by regression of expected vs. obtained values. Precision was determined by means of a 4-day validation exercise where samples at three concentration levels (1000, 100 and 10µM for model mixtures, and 200, 150, 75µM for gut-model samples) were measured in duplicate[5,6]. Repeatability (or r) represents the reliability of the quantification for duplicate analysis realised in a short period of time (the same day) for a given analyst in a given laboratory, when reproducibility (or R) concerns longer periods of time, a one day interval in this case, for the same analyst in the same laboratory. Repeatability and reproducibility were calculated using ANalysis Of VAriances (ANOVA) and the relative standard deviations (RSD_r and RSD_R) are obtained by dividing respectively r and R by 2.8 and by the concentration level. Limit of Detection (LoD) and Quantification (LoQ) were determined by taking respectively 3 and 5 times the standard deviation from a tenfold measurement of a sample with a concentration near the expected LoD.

3 RESULTS AND DISCUSSION

3.1 Detection Limits

For the chosen measurement conditions (64 scans, 600 MHz, cryoprobe, 3 mm tubes) the Limits of Detection (LoD) and Quantification (LoQ) were respectively 3 and 10 µM for model mixtures dissolved in CD_3OD and D_2O. For samples taken from in-vitro gut models LoD and LoQ were respectively 10 and 30 µM. Although the latter detection limits are less favourable as for model mixtures, these are still fit-for-purpose for rapid quantification of typical in vitro gut model samples. For more dilute samples sub-µM LoD/LoQ's should still be feasible within reasonable NMR measurement times (<1h).

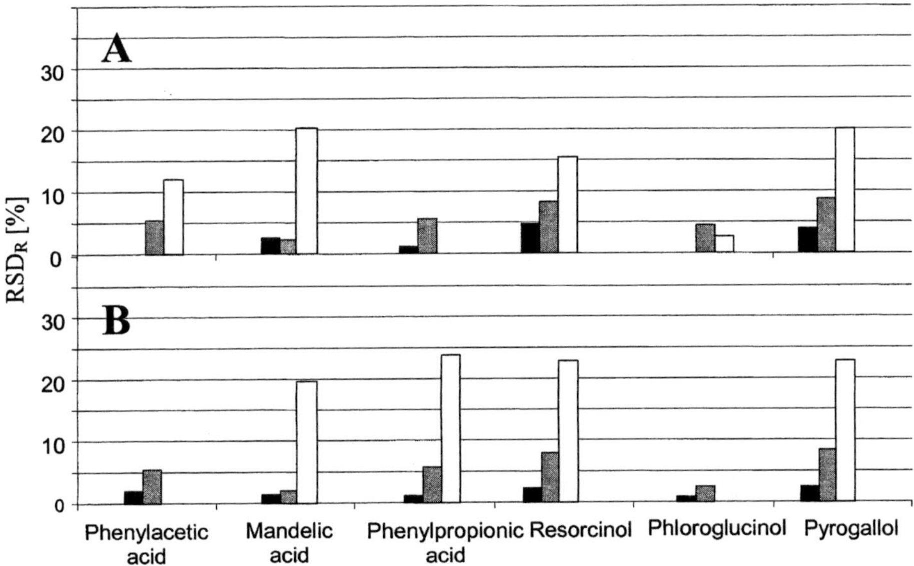

Figure 1. *Relative standard deviations of reproducibility (RSD$_R$) of phenolic acid levels (1000 (black), 100 (grey), 10 (white) µM) for IS (A) and ES (B) methods in CD$_3$OD. NMR measurements performed at 600 MHz, 3 mm SampleJet tubes, 5 mm cryoprobe and 64 scans. Missing values correspond to cases where within-day variation was larger than between-day variation.*

3.2 Precision

Model mixtures of 14 phenolic acids were prepared at three different concentration levels (10, 100, 1000 µM) in both CD_3OD and D_2O. The mixtures were measured by NMR in duplicate during 4 days and phenolic acids were quantified by means of Internal (IS) and External Standards (ES). Repeatability (r) and reproducibility (R) were determined by means of ANOVA. No significant difference between r and R was found, indicating that between-day variation is of similar magnitude as between-day variation. Typical RSD$_R$ values of phenolic acids levels as determined by IS and ES methods are presented in Figure 1. Note that in several cases RSD$_R$ could not be reported due larger variation between duplicates than between days. This typically occurs for lower concentrations where within-day variation is so large that it cannot be discerned from between-day variation. Both methods perform in a similar manner with respect to precision, with RSD_R ranging between 1-20% for levels in the mM-µM region. Note that at LoQ levels (10 µM) still fair RSD_R values of app. 20% is achieved.

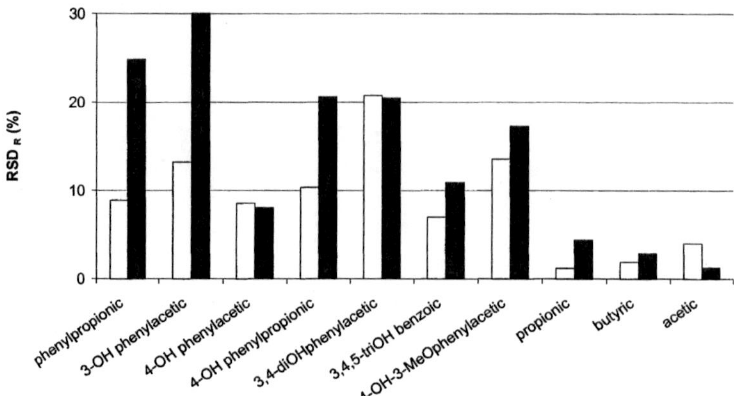

Figure 2. *Reproducibility (RSD_R) of short chain fatty acid (SCFA, 10, 1 mM, HL, LL) and phenolic acid levels (200 (white), 70 (black) µM). NMR measurements were performed at 600 MHz, 3 mm SampleJet tubes, 5 mm cryoprobe and 64 scans.*

Subsequently, method precision was assessed for real-life samples taken from in-vitro gut models that were spiked with known amounts of phenolic acids (70, 100 µM). Results are shown in Figure 2 and show that RSD_R ranges between 10-20%. Hence, precision is less favourable for these samples than for the model mixtures dissolved in CD_3OD and D_2O. We attributed this to the larger compositional complexity of these samples and the increased line width of the NMR signals. Within Figure 2 also RSD_R values are shown for short-chain fatty acids (SCFA's), which were present at much higher levels (1-10 mM). Here precision (RSD_R) at percentage level can be observed.

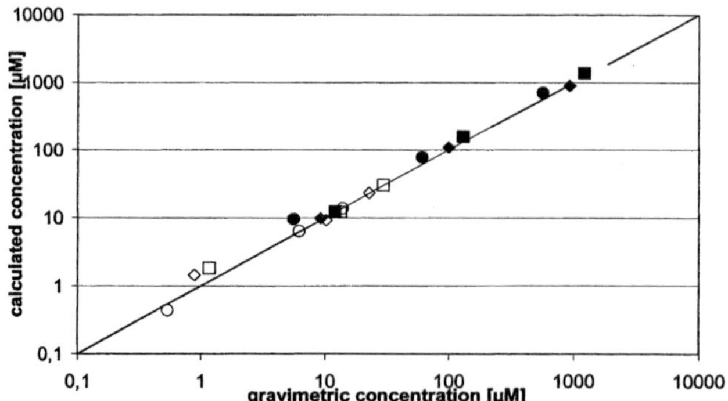

Figure 3 *Agreement between concentration levels in model mixtures in methanol as obtained by NMR (ES (closed) and GCMS (open) over the different operational ranges of both methods (○,3,4-OH phenylacetic acid; □, 3-OH phenylacetic acid; ♦, 4-OH phenylacetic acid). NMR measurements were performed at 600 MHz, using a 3 mm SampleJet tubes, a 5 mm cryoprobe and 64 scans.*

3.3 Accuracy

Model mixtures at different concentration levels (1-1000 μM) were measured by NMR and GCMS. Both methods showed good agreement (Figure 3) with gravimetric values in their respective operational ranges (NMR: 10-1000 μM, GC: 0,1-10 μM). Hence NMR can seamlessly be deployed in the concentration range where GC detectors typically show saturation.

4 CONCLUSION

Quantification by means of IS and ES methods provided accurate results within the μM-mM range for phenolic acid model mixtures dissolved in CD_3OD and D2O. Limits of Quantification is 10 μM level for a modest measurement time (64 scans) on a cryoprobe with limited sample volume (300 μM SampleJet tube). Precision was similar for IS and ES methods and ranged between 1-20% for levels in the mM-μM region. The IS and ES approaches were successfully tested and validated for quantification of gut microbial bioconversion products of polyphenols in aqueous in-vitro colon model samples.

Acknowledgements

Part of this project was carried out within the research program of the Netherlands Metabolomics Centre, which is part of the Netherlands Genomics Initiative/Netherlands Organization for Scientific Research.

References

[1] G. Gross, D. Jacobs, S. Peters, S. Possemiers, J. van Duynhoven, E. Vaughan, T. Van de Wiele, *Journal of Agriculture and Food Science* 2010, *58* **(18)**, 10236–10246.

[2] J. van Duynhoven, E.E. Vaughan, D. Jacobs, R. Kemperman, E,J.J. van Velzen, G. Gross, L. Roger, S.Possemiers, A.K. Smilde, J. Doré, J.A. Westerhuis, and T. Van de Wiele, *PNAS*, online doi: 10.1073/pnas.1000098107.

[3] I.W. Burton, M.A. Quilliam and J.A. Walter, *Analytical Chemistry, 2005*, **77(10)**, 3123-3131.

[4] C. Grun, F. van Dorsten, D. Jacobs, E. van Velzen, M. Bingham, H.G. Janssen, J. van Duynhoven *J.Chrom.B* 2008, **871**, 212–219

[5] ISO 5725, Accuracy (trueness and precision) of measurements, methods and results, Part 1–4, 6 (1994). 372

[6] W. Horwitz, *Pure Appl. Chem.* 1995, **67**, 331–343

ELUCIDATION OF NON-ENZYMATIC BROWNING REACTION PATHWAYS BY MEANS OF THE CARBON-BOND LABELING TECHNIQUE

M. Ilse[1], O. Frank[1] and T. Hofmann[1]

[1] Chair of Food Chemistry and Molecular Sensory Science, Technische Universität München, Lise-Meitner-Str. 34, 85354 Freising, Germany

1 INTRODUCTION

Besides aroma and taste, the brown coloration is one of the most important attributes of the Maillard reaction between reducing carbohydrates and amino acids. However, due to the complexity of this reaction and the multiplicity of reaction products formed, little is known about the chemical nature and the formation pathways of these nonenzymatic browning products. Therefore, it seems to be a promising approach to study suitable model systems for obtaining detailed information on the nature of the chromophoric compounds and their formation pathways involved in color formation.[1]

In order to visualize the molecular pathways in non-enzymatic browning reactions, the so called carbon-bond-labeling-technique combined with ^{13}C NMR spectroscopic analysis was applied on model browning systems.[2] This labeling strategy offers the possibility to obtain ^{13}C-^{13}C couplings and to identify which structural moieties originate from different glucose molecules. Labeling experiments in heated model reactions of glucose/alanine solutions with 5% $[^{13}C_6]$-glucose were carried out to follow the joint transfer of several ^{13}C atoms en bloc into colored compounds or precursors by LC-MS and NMR isotopomer diagnosis.[3]

2 METHODS

2.1 Maillard-Model-Reaction

A mixture of L-alanine (0.2 mol) and D-glucose (0.2 mol) was dissolved in phosphate buffer (NaH_2PO_4, 0.2 mol/l, pH = 5) and, then, refluxed for 3 h.

2.2 Labeling Strategy

The use of ubiquitary ^{13}C-labeled glucose as a precursor in the Maillard reaction will lead to universally labeled compounds, implying that individual carbon fragments involved in the formation of the colored compounds or their precursors cannot be visualized. In comparison to singly or site specific labeled carbohydrates, the advantage of multiply ^{13}C-labeled precursors is that the common transfer of several ^{13}C atoms en bloc into the target compound confirms that the bonds have remained intact during the passage of this fragment in the Maillard reaction cascade. Determining the number of carbon atoms which can be transferred en bloc would afford rigorous constrains for the intermediary pathway through which this carbon module has been processed. The site specific visualization of intact carbon bonds by measuring homonuclear ^{13}C-^{13}C coupling constants using ^{13}C spectroscopic analysis seems to be a powerful tool for structure elucidation of colored compounds and clarification of their formation pathways.

2.3 Nuclear Magnetic Resonance Spectroscopy (NMR)

The NMR-spectra were recorded on a 500 MHz Bruker Avance III Spectrometer with TCI Cryo-Probe. The ^{13}C-Spectra were measured at 125 MHz using the standard CPD sequence for proton decoupling.

3 RESULTS

Some very potent color precursors formed from the Maillard Reaction have been investigated by using this carbon-bond-labeling-technique. As an example the structure of 2-(2-formyl-3-(1,2,3-trihydroxypropyl)-1*H*-pyrrol-1-yl)propanoic acid **1** (Figure 1), could be unequivocally identified by this technique and its formation pathway could be confirmed to run via an intact C_6 carbon skeleton as well as a C_2 carbohydrate fragment by means of isotopomeric pattern analysis. The ^{13}C-^{13}C couplings clearly showed that the C_2 fragment is located at position 4 and 5, indicated by a doublet with a typical $^1J_{C,C}$ coupling constant of 62 Hz (Figure 1). The absence of further couplings at the positions 4 and 5 clearly indicates that a C_2 fragment was incorporated into the target compound.

On basis of these NMR data, the reaction pathway displayed in Figure 2, was proposed for the formation of **1**. The reaction of the Amadori product with the 3-Deoxyosone led to a trisubstituted pyrrole by elimination of two molecules of water (Figure 2). The proton catalyzed elimination of a C4-Fragment gives rise to the target compound.

4 CONCLUSION

NMR assisted bond labeling technique is a powerful tool to confirm structures of complex organic molecules and offers insights into the formation pathways on a molecular level. Information obtained by such studies is helpful to extend the knowledge on colorants generated by the puzzling network of Maillard-type reactions during food processing and will help to construct a route map of reactions leading to color development in heated food stuffs.

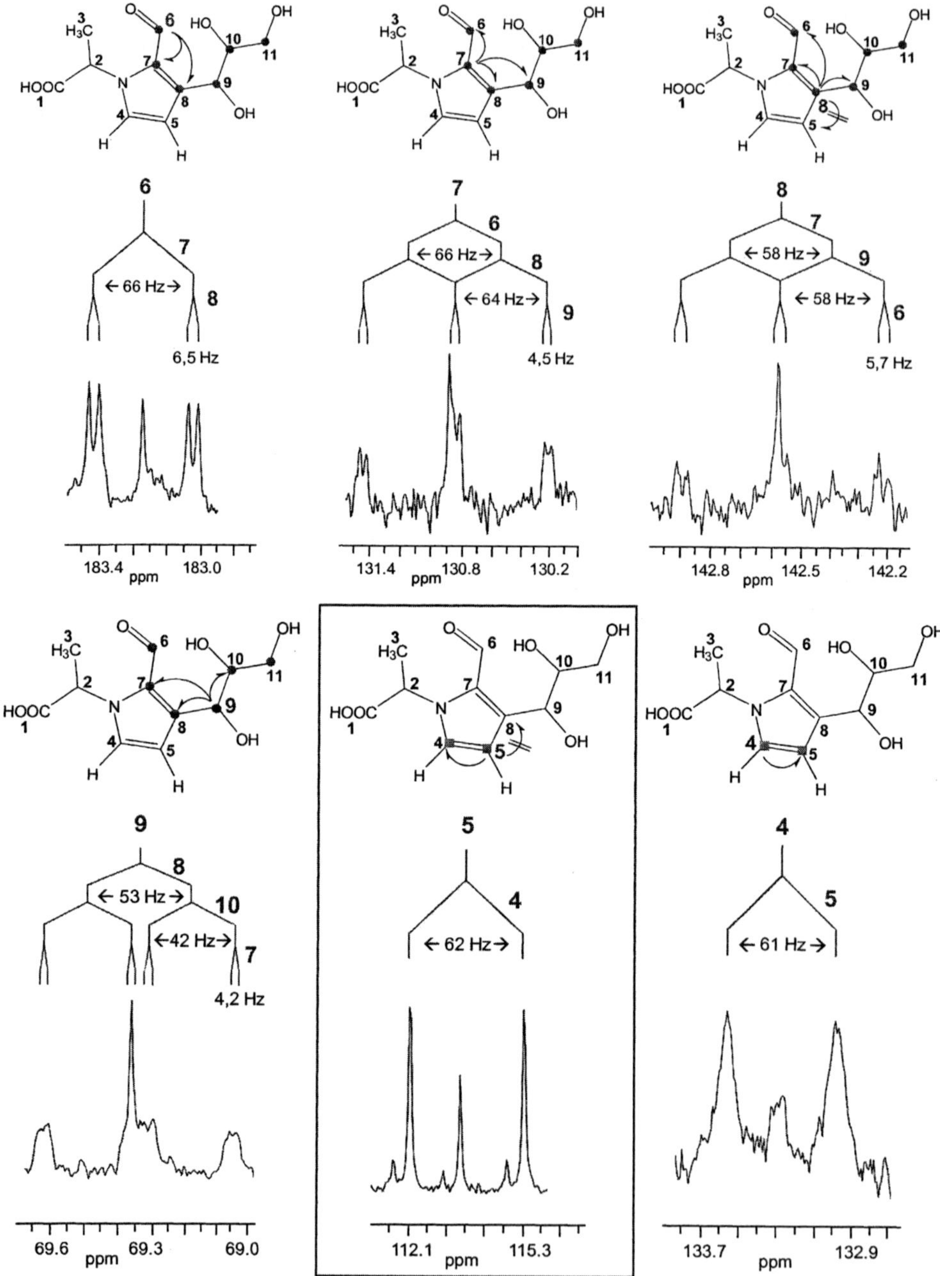

Figure 1 *Excerpts of the ^{13}C-Spectra of isotopomer **1** isolated from a heated D-glucose and L-alanine mixture containing 5% [^{13}C$_6$]-glucose.*

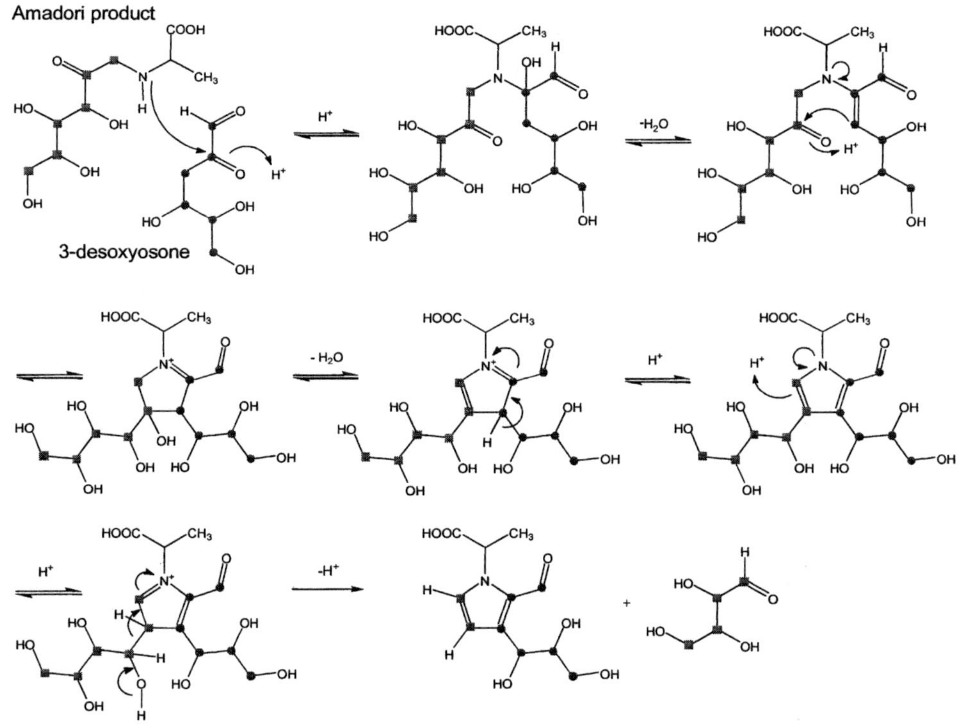

Figure 2 *Proposed reaction pathway leading to the formation of 1 from glucose and alanine (dots in the structures indicate* ^{13}C *labels;* • *and* ■ *differentiate* ^{13}C *atoms originating from different glucose molecules).*

References

1   T. Severin and U. Krönig, *Chem. Mikrobiol. Technol. Lebensm.*, 1972, **1**, 156.
2   A. Bacher, Ch. Rieder, D. Eichinger, D. Arigoni, G. Fuchs and W. Eisenreich, *FEMS Microbiol. Rev.*, 1999, **22**, 567.
3   O. Frank and T. Hofmann, *J. Agric. Food Chem.*, 2002, **50**, 6027.

COMPARATIVE STUDY OF THE THERMAL AND MICROWAVE OXIDATION IN OLIVE OIL. ^{31}P-NMR QUANTITATIVE DETERMINATION OF 1,2 AND 1,3-DIGLYCERIDES AND OTHER MINOR COMPOUNDS.

C. Lucas-Torres, M. Moreno, A. Juan, A. de la Hoz, and A. Moreno.

Department of Inorganic, Organic Chemistry and Biochemistry, University of Castilla-La Mancha, Av. Camilo José Cela 10, 13071 Ciudad Real, Spain.
E-mail: Andres.Moreno@uclm.es

1 INTRODUCTION

Virgin olive oil is the most commonly used cooking fat in Mediterranean countries. The main two techniques for food preparation are: frying and microwaving. During the heating process, due to the high temperature and absorption of oxygen and water, the 98 % of triacylglycerols[1] present in the olive oil suffer changes; also cause differences in the olive oil composition, increasing the diglyceride content, and generating primary oxidation products, whose degradation generates secondary oxidation products.[2] In fresh olive oil, the 2% remaining, concerns to minor compounds such as di- and monoglycerides (DG and MG), polyphenolics, antioxidants, pigments, etc. which could also suffer some changes when olive oil is heated.

Many studies show the importance of NMR in the olive oil field. ^{1}H and ^{13}C-NMR spectra of a fresh olive oil sample,[3] elucidate the olive oil main components. In addition, oxidized olive oil ^{1}H and ^{13}C spectra, show the main oxidized compounds as aldehydes, hydroperoxides, etc.[4].

The main drawback of the mentioned method is that minor compounds are overlapped in NMR spectra by triacylglycerols. Fresh and oxidized olive oil spectra are very similar due to this lack of information. Reagent containing phosphorus derivatized free –OH groups in olive oils' compounds[5], which allows us to obtain a ^{31}P-NMR spectrum of them. Therefore the ^{31}P-NMR is becoming the best alternative way to observe the differences between fresh and oxidized olive oils.

The aim of the present work is to compare the diglyceride composition of thermally and microwave oxidized Castilla-La Mancha's olive oils, using a facile method based on ^{31}P-NMR, and to detect other minor compounds.

2 METHOD AND RESULTS

2.1 Heating Processes

The study was carried out on ten different Castilla-La Mancha's olive oils from olive varieties as Arbequina, Cornicabra, Hojiblanca and Picual. Thermal and microwave oxidation at 190°C were performed during 4 hours[6]. Classical heating were provided with a heat-on device and a bubbling pump air connected. The microwave reactor were provided with a pump to bubble air inside the flask, operating at a of 200 watts power.

2.2 Sample Preparation and ^{31}P-NMR Spectral Acquisition

A solution, which consists of solving 0.6 mg of chromium acetylacetonate (III), Cr(acac)$_3$ (0.165 µM) and 13.5 mg of cyclohexanol (13.47 µM) in 10 mL of a mixture of pyridine and CDCl$_3$ solvents (1.6:1 volume ratio), was protected from moisture with molecular sieves[5]. Cyclohexanol is used as internal standard (I.S.) in order to quantify the different compounds after NMR analysis.

150 mg of oxidized olive oil were dissolved in 0.4 mL of the solution, and 20-45 µL of TMDP (2-chloro-4,4,5,5-tetramethyl-1,3,2-dioxaphospholane), depending on the heating treatment, were added. The mixture was left to react for half an hour (Figure 1), and then was used to acquire the ^{31}P-NMR spectra.

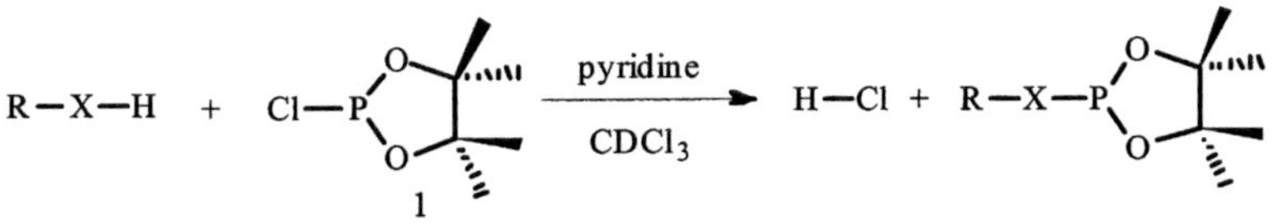

Figure 1 *Derivatization reaction of olive oil minor compounds and Reactive 1 (2-chloro-4,4,5,5-tetramethyldioxaphospholane, TMDP) in pyr:CDCl₃ solution.*

All samples were prepared and the spectra were recorded three times.

^{31}P-NMR spectra were obtained on a Varian Gemini 400MHz operating at 162.085MHz for ^{31}P nuclei, at 25°. The spectra were acquired employing the inverse gated decoupling pulse sequence in order to suppress NOE effects. Quantitative parameters were used: 90° pulse width 15µs, relaxation delay 1s, spectral width 10KHz, 128 scans. All chemical shifts is referred to the internal standard cyclohexanol shift at 145.2 ppm.

2.3 Quantitative Calculations

Using integral areas of each olive oil spectra, calculations of the quantity of minor olive oil compounds may be resolve[7] by two different ways:

$$\text{Compound}\,(\text{mg}) = \frac{I_{compound}}{I_{standard}} \cdot n\,(\text{mmol standard}) \cdot M_w\,(\text{compound})^5 \tag{1}$$

$$\%\,\text{compound}\,(wt) = \frac{{I_{compound}}/{N_{compound}}}{\sum_{i=1}^{m} {I_i}/{N_i}} \cdot 100^7 \tag{2}$$

Equation 1 for calculates the amount of each interesting compound (mg).
Equation 2 for calculates the compounds' percentage.

^{31}P-NMR spectra of a fresh, thermal and microwave oxidized olive oils shows the main differences in its compounds, as illustrates the figure 2.

The main compounds identified in 31P-NMR spectra were diglycerides and free fatty acids. Thermal oxidation shows more oxidized compounds than microwave oxidation. Thermal oxidation furthered rise 1,3-diglyceride respect to 1,2-diglyceride. Also, the fatty acid content is bigger in thermal oxidation than in microwave oxidation. In addition, the β-sitosterol content disappears when olive oil is heated. The chemical shifts of the principal compounds observed are shown in Table 1.

Table 1   *Main Chemical Shifts (ppm) of Olive Oil when ^{31}P-NMR is applied to fresh and oxidized samples.*

Signal	Chemical Shift (ppm)	Assignment
a	174.89	Free TMDP
b	148.20	1,2-DG
c	146.67	1,3-DG
P. I.	145.20	Cyclohexanol (I.S.)
d	144.90	β-sitosterol
e	134.77	Free Fatty Acids
f	132.18	Product from reaction of TMDP with water

The measure of integral areas allow us calculate the percentage of 1,2 and 1,3 diglyceride contents, which are shown in Table 2.

Table 2   *% 1,2 and 1,3-diglycerides calculate from integral values of ^{31}P-NMR spectra*

	Hojiblanca		Cornicabra		Arbequina		Coupage (Cornicabra, Picual and Arbequina)	
	1,2-DG	1,3-DG	1,2-DG	1,3-DG	1,2-DG	1,3-DG	1,2-DG	1,3-DG
Fresh Olive Oil	0.91	0.07	0.85	0.16	1.01	0.22	1.01	0.14
Thermal Oxidation	0.86	1.54	1.95	3.58	1.83	3.25	1.94	2.85
Microwave Oxidation	0.37	0.72	0.37	0.74	0.45	0.80	0.29	0.49

A relation between total diglyceride content and the oxidation level in an olive oil is observed. Thermal oxidation provides more total diglyceride percentage than microwave oxidation. This may be observed because thermal oxidation produces a pronounced degradation of triglycerides than microwave oxidation. In fact, the latter oxidative process shows a total diglyceride percentage very similar to the percentage observed in fresh olive oil, meaning that microwave only produces isomerization between 1,2 and 1,3 diglycerides.

The graphical observation of these results and the comparison between fresh and the two types of oxidized olive oil is represented in Figure 3.

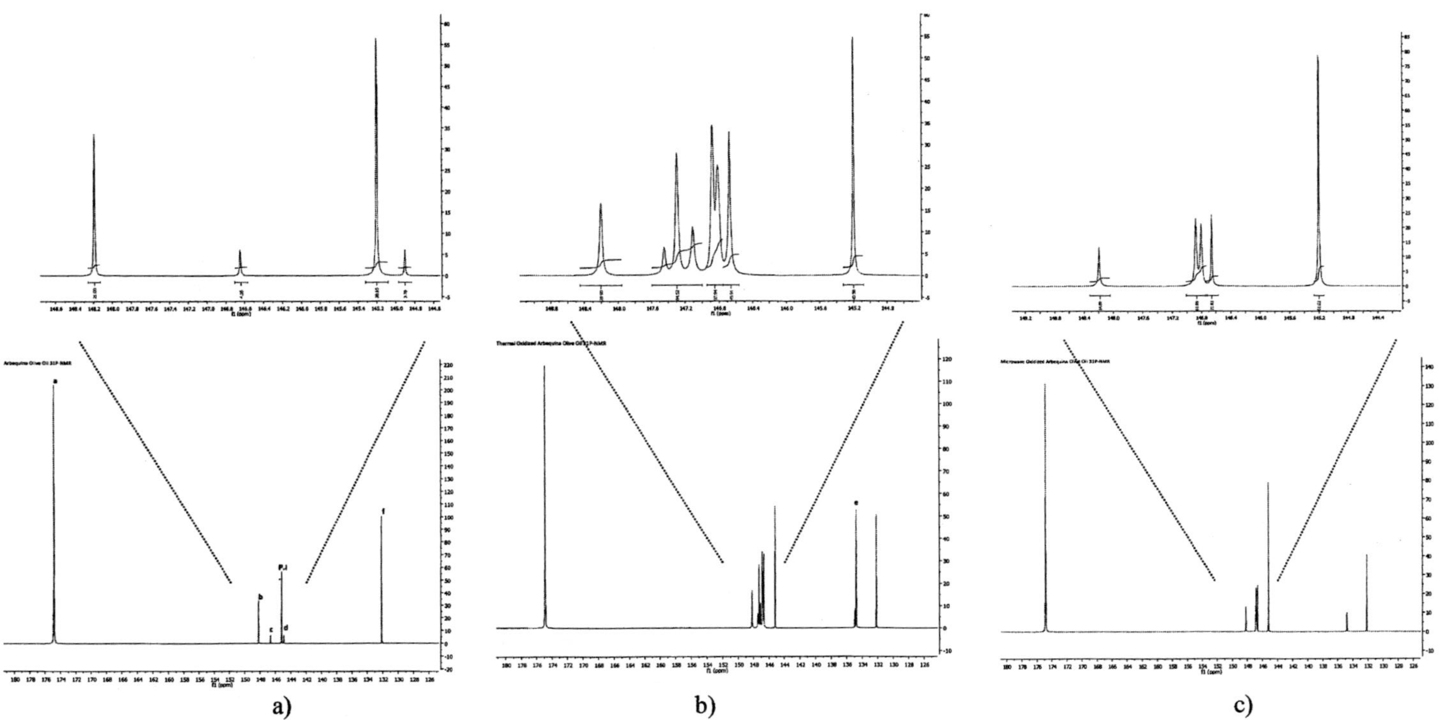

Figure 2 ^{31}*P-NMR spectra of Arbequina Olive Oil: a) Fresh Olive Oil; b) Thermal Oxidized Olive Oil; c) Microwave Oxidized Olive Oil. The expanded zone of each spectrum show the diglycerides signals and other peaks not identified yet. Integral values are indicated below each peak.*

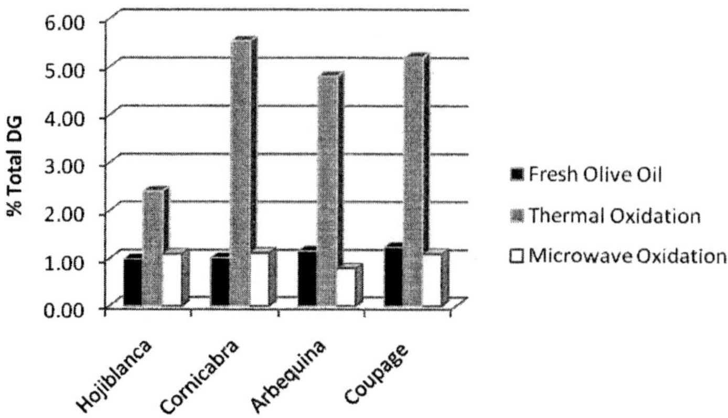

Figure 3 *Total diglyceride content of four different olive oil varieties, comparing the fresh (black) with the thermal (grey) and the microwave oxidized (white).*

3 CONCLUSION

This study demonstrates the efficiency of the ^{31}P-NMR technique to detect and quantify DG and other minor components in olive oil without extraction or separation methods before the analysis, providing a new way to detect oxidized olive oils. Also, important differences between thermal and microwave oxidation were found, in order to apply the study to the quotidian chemistry.

There have been observed changes between 1,2 and 1,3-diglyceride content depending on the oxidation treatment. Concentration of 1,2-diglyceride decreases in favor of 1,3-diglyceride concentration when high temperature is applied to olive oil. The total diglyceride contents would be used to evaluate the differences between the two oxidative processes. Thus, microwave oxidation only produces isomerization between 1,2 and 1,3 diglycerides, but, when thermal oxidation is applied, triglyceride degradation is observed.

References

1   M.T. Morales and M. Leon-Camacho, 'Liquid and Gas Chromatography: Methodology applied to olive oil' in *Handbook of olive oil*, eds., J.L. Harwood and R. Aparicio, Aspen Publishers, Inc., Maryland, 2000, Chapter 7, p. 160.

2   E.N. Frankel, 'Hydroperoxide formation' and 'Hydroperoxide decomposition' in *Lipid oxidation*, The Oily Press, University of California, 1998, Chapter 2 and 3, pp. 23-77.

3   R. Sacchi, F. Addeo and L. Paolillo, *Magn. Res. Chem.* 1997, **35**, 133-145.

4   M.D. Guillén and A. Ruiz, *Food Chem.* 2006, **96(4)**, 665-674.

5   E. Hatzakis, G. Dagounakis, A. Agiomyrgianaki and P. Dais, *Food Chem.* 2010, **122(1)**, 346-352.

6   F. Caponio, A. Pasqualone and T. Gomes, *Eur. Food Res. Technol.* 2002, **215**, 114-117.

7   U. Holzgrabe, *Prog. Nucl. Mag. Res. Sp.* 2010, **57**, 229-240.

TIME DOMAIN ^{1}H-NMR OF ARABINOXYLANS AND 1 β-GLUCANS FILMS, MODELS OF A LAMELLAR ORGANISATION IN ENDOSPERM CELL WALLS OF CEREAL GRAINS

R. Ying, J. Ruellet, C. Rondeau-Mouro[#]

INRA, UR 1268 BIA, Equipe Parois Végétales Polysaccharides Pariétaux, BP 71627, F- 44316 Nantes, France
INRA, UR 1268 BIA, Plateforme BIBS, BP 71627, F-44316 Nantes, France

1 INTRODUCTION

Plant cell walls are partly responsible for the variability of plant crops in their different agro-industrial food and non-food uses. Control of this variability is essential for the improvement of plant productions quality. Endosperm cell walls have important impact on the uses of cereal grains (milling, baking, human nutrition…)[1-2]. Arabinoxylans (AX) and mixed (1-3)(1-4)-β-D-glucans (BG) are the main components of the cereal grains endosperm cell walls[1,3-4]. They constitute respectively 70 and 30% of the walls, which represents 2-4% of the dry matter in endosperm. AX are characterized by a linear backbone of xylose to which arabinose substituents are attached while BG are linear homopolymers of D-glucose linked by β-(1-4) and β-(1-3) linkages. The arabinofuranose units are essentially found as mono- or di-substitution in position O-3 or O-3 and O-2 of xylose residues[1,5]. AX exhibit large natural variations on their structure, which mainly differ by the arabinose to xylose ratio (A/X) with values ranging from 0.31 to 1.06 [6-7]. Arabinosyl residues can also be esterified on the O-5 position mainly with ferulic acid[9-10]. The local composition heterogeneity of endosperm cell walls has been investigated on samples recovered from fractioning processes[11], by direct detection of fluorescence[12], by coupling imaging and spectroscopic techniques[13-15] and by immunolabelling techniques[16-17]. It was concluded that BG are present in higher proportion in cell walls of aleurone and subaleurone layers than in the starchy endosperm, which contains in its central zone, a higher content of AX. Aleurone cell walls are characterised by a low A/X ratio in the range 0.3-0.4, while it attains 0.5-0.6 in the starch endosperm walls. This chemical heterogeneity has been extensively described but little is known about the time-course and the pattern of the AX and BG deposition in cell walls during endosperm development. Various observations using different microscopy techniques (MCBL, AFM) have indicated the lamellar organisation of wheat endosperm cells walls that possibly reflects the assemblies of AX and BG[17]. In order to study the impact of their fine structure and interactions on cell walls properties, in particular hydration properties, we have prepared AX and BG films as models of their lamellar distribution within endosperm cell walls. The second moment M2 and spin-spin relaxation times T2 were measured respectively with single pulse free induction decay (FID) and Carr-Purcell-Meiboom-Gill (CPMG) sequences. The objectives of this work were to study the influence of both water and temperature on mobility and hydration of AX and BG within films, in relation to their organization 1 in cell-walls of wheat grains.

2 MATERIALS AND METHODS

2.1 Materials

Water-extractable arabinoxylans were prepared from wheat flours according to the method of Dervilly-Pinel et al.[7]. Their purification and characterization have followed the procedures described in Dervilly-Pinel et al.[7] and have been applied to the Barley water soluble β-glucans (medium viscosity) purchased from Megazyme.

The A/X ratio, found to be 0.33 was conforming already published results concerning the fractionation with ethanol of arabinoxylans[7,8]. AX-0.33 contained arabinose and xylose with 0.23% (% p/p) ferulic acid. The macromolecular characteristics of each polysaccharides were determined using the methods described in Dervilly-Pinel et al.[7,8]. AX-0.33 and BG were characterized (respectively) by a weight-average molar mass of 177 10^{-3} g/mol and 218.4 10^{-3} g/mol, a radius of gyration of 38 nm and 14 nm, a similar polydispersity index of 1.5 and an intrinsic viscosity of 295.6 and 331.2 ml/g. The characteristics for AX are similar to values obtained by other authors[7,8].

2.2 Film Preparation

The AX and BG solutions were cast into polystyrene Petri dishes (PS, Ø 5 cm). About 20 mg of arabinoxylans was used for each film (film thickness 10 μm, Ø 5 cm). The films were dried in a climate room at 40 °C and 40% RH.

2.3 Sorption isotherm

The experimental water sorption isotherm of the films was determined at 20 °C using the saturated salt method. Films were equilibrated at various water activities in desiccators containing saturated salt solutions of known relative humidity RH : 11% (LiCl), 59% (NaBr)), 75% (NaCl), and 91% (BaCl2). The sorption of water was followed gravimetrically until equilibrium was achieved (generally within 15 days).

The glass transition temperature T_g of AX and BG films were measured by a calorimetric DSC Q100 differential scanning calorimeter (TA Instrument), previously calibrated with Indium. An amount of 20 mg of each sample with the adjusted moisture content considered in this work was first heated from -40 to 120 °C (60 °C/min) and second heated from -40 to 150 °C using an empty pan as a reference. The glass transition temperature of the specimens was determined from the midpoint of the heat capacity change observed on the second scan, at a heating rate of 3 °C/min.

2.4 NMR spectroscopy

^{1}H NMR measurements were performed using a low-field NMR spectrometer (The Minispec, Bruker SA, Wissembourg, France) operating at a resonance frequency of 20 MHz for [1]H (0.47T). The NMR system was equipped with a temperature control device connected to a calibrated optic fiber (Neoptix Inc., Canada) allowing a ± 0.1 °C temperature regulation. The sample coil was 8 mm in diameter. Tubes were filled to about 10 mm in height in order to place samples in a homogeneous region of the NMR magnet, then weighed and hermetically closed. A Teflon rod was introduced and filled the dead volume of each tube to avoid water loss. Only a hole of 1.2 mm in diameter permitted to pass the optic fiber through the cap and the teflon rod to measure temperature of the sample. Thermal equilibration was ensured by allowing a 7 min-waiting time after each temperature step before the experiment was started. The dead time of the ^{1}H NMR probe was 11μs. Proton free induction decays (FID) were acquired using the following parameters: a 90° pulse of 3.4 μs, a dwell time of 0.5 μs between two successive data

points, 160 scans of 19900 data points and a recycle delay of 2 s between each scan. The spin-spin relaxation (T2) was measured from Carr-Purcell-Meiboom-Gill (CPMG) pulse sequence. The delay between the 90° and 180° pulses of the CPMG sequence was 40 μs. 160 scans were acquired with 800 data points and a recycle delay of 2 s between each scan.

2.5 Data analyses

Spin-spin relaxation data were analysed with the following model:

$$I_{CPMG}(t) = \sum_{i=i}^{i} A_i \times exp\left(-\frac{t}{T_{2i}}\right) \qquad \text{Equation 1}$$

The curves were then fitted with two different methods: the discrete method Marquardt [18] and the CONTIN method[19]. As the results given by both methods were consistent, only the results from the Marquardt method are presented here.

The second moment M_2 values were determined from the analysis of the beat or oscillation observed in FID. The FID were fitted to the following equation which is a combination of a sinc function and a Gaussian broadening[20] :

$$I_{FID}(t) = A \exp\left(-\frac{a^2.t^2}{2}\right).\frac{\sin bt}{bt} + B \exp\left(-\frac{t}{T_2^*}\right) \qquad \text{Equation 2}$$

In this equation, the parameters A and B represent the contributions of the immobile and mobile protons, while T_2* denotes the spin-spin relaxation time of the mobile proton fraction in presence of the magnetic field inhomogeneities. The NMR spectrum of the immobile proton fraction is assumed to be a rectangular line shape with a total width $2b$, convoluted with a Gaussian line shape with a standard deviation given by parameter a.[20-22] The second moment M_2 of the broad line shape, which is a measure of the strength of the dipolar interactions, is given by[21-22] :

$$M_2 = a^2 + \frac{1}{3} b^2 \qquad \text{Equation 3}$$

3 RESULTS AND DISCUSSION

Study of the free-induction decay (FID) observed for AX and BG films indicated an oscillation (or a beat) of the NMR signal arising from residual local order within the samples (Figure 1). This short-range organisation induces between protons strong dipolar interactions, characterized by the second moment M_2. The sinusoidal pattern [1] observed on FID has already been depicted in glassy oligosaccharides as maltose[21-22] or starch[23-24]. As the water content or temperature of samples increased, the FID beat pattern was less pronounced as a consequence of a decrease of the number and/or strength of the dipolar interactions within AX and BG films (figure 1). This phenomenon could result from (i) a reduction of the proton density (ii) an increase of the proton distances and/or (iii) an improvement of the anisotropic mobility of the polysaccharides chains which in the same time diffuse more rapidly.

The additional slow decaying component of the FID signal was attributed to the mobile protons of water whose content increased with the relative humidity (RH) of samples.

Some authors have determined that in presence of increasing amount of water, the signal of the mobile protons display a faster decaying in relation to higher water mobility [22,25]. In order to describe this process, spin-spin relaxation times T2 of water have been measured within the samples using the CPMG sequence.

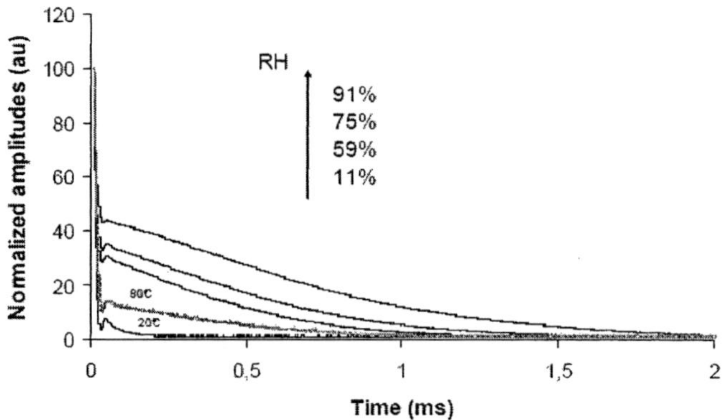

Figure 1 *FID signals of BG films at different water contents (see Table 1 for RH between 17 11 and 91%) measured at 20°C (continuous line) and 80°C at RH=11% (dotted line).*

Table 1 *Values of T2TP, TTP, Ea and Tg for BG and AX films and for the BG powder at different water contents.*

Sample[a]	% water content (RH)[b]	TTP (°C)	T2TP (ms)	Ea (kJ.mol⁻¹)	Tg (°C)
AX DF	4,17 (11)	> 80	nd	40,4	> 150
AX DF	15,24 (59)	75	7,80	38,2	82,9
AX DF	17,58 (75)	60	8,98	33,9	58,7
AX DF	28,31 (91)	32.5	12,05	30,8	nd
BG DF	4,71 (11)	> 80	nd	34,9	> 150
BG PW	16,43 (59)	70	6,01	30,2	nd
BG MF	16.43 (59)	65	4,70	31,9	nd
BG DF	16,43 (59)	65	3,62	29,4	101,55
BG MF	19,40 (75)	45	4,93	19,6	nd
BG DF	19,40 (75)	55	3,65	21,3	70,94
BG DF	28,86 (91)	35	4,47	20,8	29,11

[a] DF refers to stacked discs of films, PW for powder, MF for minced films. T2TP is maximum value of T2 for the corresponding temperature TTP. Ea is the energy of activation. [b]21 Water contents measured at 20°C.

Three forms of samples containing BG have first been investigated in order 1 to evaluate the impact of the polysaccharides organisation on T2 and M2 values. Then comparison between BG and AX (A/X=0.33) films has been carried out, aimed at investigating some eventual hydration and mesostructure differences in relation to their organisation within endosperm cell walls.

3.1 Impact of the sample form on the NMR measurements

Various preparations of samples were analyzed in order to study the impact of their form on NMR measurements. Three preparations were tested: powder, minced films and discs of films stacked inside the tube. These measures were conducted on BG samples available in large quantities in the laboratory.

The second moment M_2 as a function of temperature showed very little variation between the three forms of samples. Indeed, for a 16.4% water content, the M_2 varied between 7921 $10^6.rad^2.s^{-2}$ and 5139 $10^6 rad^2.s^{-2}$ for temperatures between -40 ° C and +80 ° C, with variations that did not exceed 400 $10^6. rad^2.s^{-2}$ between the three samples (figure 2A). By increasing the water content at 19.4%, not only the M_2 values decreased but also differences between minced films and discs of films are more pronounced.

Comparison of the proton spin-spin relaxation times T_2 for each of these samples was another manner to attest the importance of the sample preparation. Different previous works have referred to spin-spin T_2 relaxation time measures to study hydration phenomena in biological systems[26]. In plants, the multi-exponential behavior of water has been connected to different mobilities of water that can interact with solid phases in exchange with water molecules in various compartments of various sizes (intracellular water, extracellular water, water in vacuole ...)[26]. In the present work, the water was expected to interact with polysaccharides which are molecules strongly hydroxylated. Rapid chemical exchanges (of the order of the picoseconds) should take place between the so-called "free water" located at the surface of films and/or within the film mesopores and "bound water" according to this simplified reaction :

$$H\text{-}O\text{-}H...HO\text{-}P* \leftrightarrow H_2O + HO\text{-}P \quad (* polysaccharides .... hydrogen bond)$$

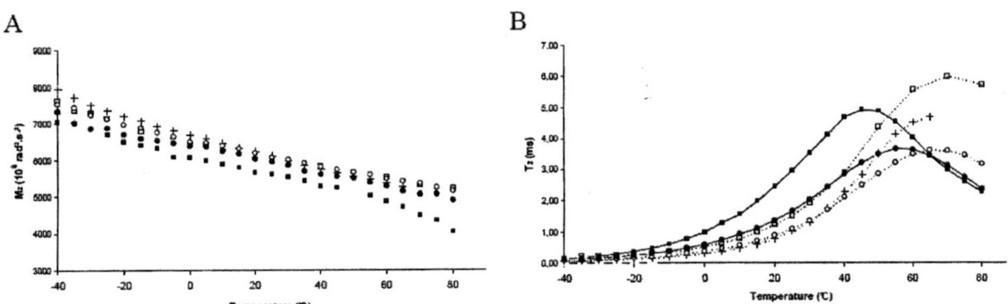

Figure 2 *M₂ (A) and T₂ (B) as a function of temperature for vrious sample form of BG 34 □ powder RH=59%; + minced films RH=59%, ○ discs of films RH=59%, ■ minced films RH=75%, ● discs of films RH=75%,*

Plot of T_2 as a function of temperature indicated constant and low values 1 (on the order of 0.1 to 0.8 ms entre -40°C and 0°C (figure 2B). The T_2 values increased from 0 °C to achieve a maximum value (T_2 on top of the peak noted T_{2TP}) at a temperature T_{TP} which varies depending on the nature, the form and hydration of samples. At higher temperatures, there is a gradual reduction of T_2 that can be explained by additional molecular diffusion process and/or averaging due to rapid motions of the polysaccharides motions (measurements under progress).

Evolutions of the T_2 values as a function of the temperature have permitted to assess the activation energy required to shift the chemical exchanges observed between the various phases of water. Between 5 ° C to 60 ° C, the $\ln(T_2)$ showed a linear dependence on the inverse of temperature. This result was consistent with growing mobility of water according to the phenomenological law of Arrhenius, which traduces the dependence of the reaction rate with temperature. The Arrhenius plot of the relaxation time dependence with temperature ($\ln(T_2)$ as a function of $1/T$) allowed calculation of the energy of activation Ea (slope of the straight line $\ln(T2)= f(1/T))$[27].

Table 1 gathers the estimated T_{2TP}, T_{TP}, Ea and T_g for BG and AX films and for the BG powder at different water contents.

Results of T_2 for the different sample forms of BG containing 16.4% of water, indicated a decreasing of the T_{2TP} values (from 6.01 ms to 3.62 ms for powder and discs of films, respectively) while few changes of T_{TP} and Ea values were observed. The activation energy values, close to 30 kJ/mol, whatever the form of sample, were in agreement with the activation energy for moisture diffusion values around 28 kJ/mol reported for various food materials[28-30]. Further experiments are under progress in order to measure the water diffusion within AX and BG films.

The increase of the moisture in BG films (at 19.4%) has slightly increased the value of T_{2TP} but for lower temperatures T_{TP} of 10 or 20 ° C : 4,93 ms at 45 °C and 3,65 ms at 55 °C for the minced films and the discs of films respectively. Activation energies are little different between the two forms of sample at RH=75% (19.6 and 21.3 kJ/mol) but decreased compared to the 59% relative humidity, consistent with faster water diffusion. These results indicate that more the sample is organized (if we consider discs of film being the most organized sample and powder the less organized), lower is the T_{2TP} value for temperature and almost equivalent values of Ea. Increasing moisture films, T_{2TP} varies little, but their temperature decreases 10-20 ° C consistent with the idea that the system needs less thermal energy to shift the water exchange equilibrium (between the "bound" and "free" states) toward "free" phases containing water diffusing more rapidly.

3.2 Dynamics of polysaccharides and water protons within films

Hydration properties within films are of major interest for the understanding of the state and the role of water against the mesostructure of AX and BG films. Eight different compositions and moisture films were analyzed : four samples of AX (A/X = 0.33) with a water content from 4.2 to 28.3 % and four samples of BG containing 4.7 to 28.9% of water (for RH = 11, 59, 75 and 91%).

3.2.1 The second moment M2 of AX and BG films
The plot of the second moment M_2 values as a function of temperature (figure 3A) showed a linear decay except for high hydrated AX films (RH= 75 and 91%) and BG films at RH=91% that displayed a slope breakdown at temperatures close to the polysaccharide glass transition temperature T_g (Table 1). For the AX samples, M_2 values 1 varied from 10455 10^{-6} rad^2.s^{-2} (-40°C) to 483 10^{-6} rad^2.s^{-2} (+80°C) while for BG films M_2 data changed from 9945 10^{-6} rad^2.s^{-2} (-40°C) to 2608 10^{-6} rad^2.s^{-2} (+80°C). For low hydrated samples (RH=11 and 59 %), second moment M_2 of AX and BG are close and fluctuated within the same range as the temperature was raised. A higher moisture within films induced more pronounced discrepancies beyond 10°C with an important decreasing of the M_2 values for AX. The results indicate that for the two types of polysaccharides, seconds moments M_2 diminished as the temperature and moisture were increased. As the proton density of the solid and mobile phases has been found unchanged against temperature (constant values of

B/(A+B) from the fitting of equations 2), two phenomena could explain a M_2 decreasing associated with weaker dipolar interactions : (i) an increase of the proton distances and/or (ii) an improvement of the anisotropic mobility of the polysaccharides chains which in the same time diffuse more rapidly. These two hypotheses are valuable in our case. In particular, an interproton distance gap can be explained regarding the polysaccharides structure. Compared to the smooth chains of BG, AX are branched polymers of xylose containing in the present work, 33% of arabinose substituents which are bound to xylose residues in position O-3 or O-3 and O-2 [1,5]. The sterical hindrance of the arabinose branch points can assess larger interproton distances which could enlarge as the anisotropic mobility of the polysaccharides increased with a rise of the temperature or moisture. Moreover, the interactions between the AX chains within films should be weaker than between BG chains. Indeed, the arabinose residues should interact strongly and preferentially with xylose residues than with other arabinose substituents of close AX chains. Stronger interactions between the BG chains were attested by the DSC measurements which indicated higher glass transition temperature T_g for BG films at RH between 11% and 91% (Table 1).The strong effects of the water content on the glass transition temperatures were also demonstrated for AX and BG films.

A   B

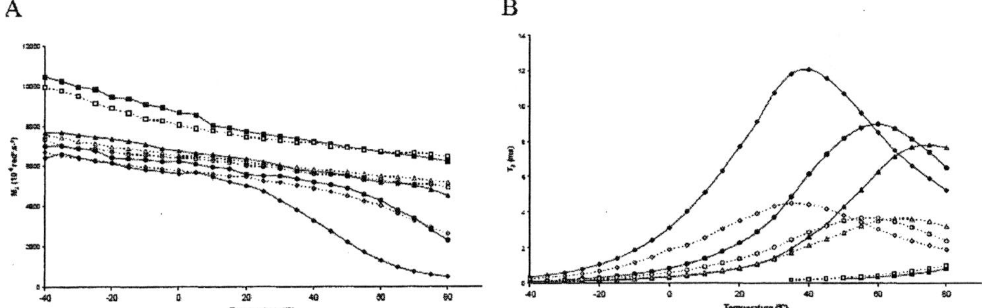

Figure 3 *M_2 (A) and T_2 (B) as a function of temperature for discs of films made of AX (filled symbols and continuous line) and BG (empty symbols and dotted line)*
□ RH=11%; △ RH=59%, ○ RH=75%, ◊ RH=91%

3.2.2 The spin-spin relaxation times T_2 of water protons within AX and BG films

The spin-spin relaxation times T_2 for the water protons were plotted as a function of temperature in figure 3B. Net differences were observed between the maximum values of T_2 (T_{2TP}) for AX and BG. As mentioned in table 1, T_{2TP} values for AX varied from 7.80 ms to 12.05 ms while for BG the T_{2TP} did not exceed 4.47 ms (for stacked discs 1 of films noted DF) depending on the water content of samples. For samples prepared in the same form, the activation energy data for BG films are lower than for AX (reduction of 3 to 10 KJ/mol- see table 1). By comparing results obtained for AX with BG, it seemed that the interactions between water molecules and the hydroxylated functions of polysaccharides were most important for the AX films (higher values of Ea and T_{SP}). However it should be noted that all the samples of AX are less moisturized than the BG samples (less "free" water within the AX films). Furthermore, chains of AX as well as water are more mobile within AX films (weaker M_2 and higher T_2 values, respectively). The differences between the activation energy calculated for AX films compared to BG films were therefore exclusively related to differences in contents of water. Nevertheless the discrepancies between the T_{2SP} values were significant and could be related to a better organization and/or more compact mesostructure of the BG films, resulting in a reduction of the water as well as polysaccharides dynamics. These hypotheses were consistent with higher second moment M_2 and glass transition temperature T_g values for hydrated BG films at temperatures beyond 10°C.

4 CONCLUSION

Various measurements carried out in this work permitted to emphasize some relations between water dynamics and the organisation and/or mesostructure of films (10 μm of thickness) made of arabinoxylans and β-glucans, polysaccharides extracted from cereal endosperm cell walls. The water mobility has been found a very complex process whose interpretation would not have been possible without the study of its dependence on the temperature and the water content within polysaccharide films. Temperature dependence of the water spin-spin relaxation times T_2 has been described by Arrhenius-type relationship. This Arrhenius relation permitted to estimate activation energy values that varied with the polysaccharide type and the water content of films. Values of Ea around 30 kJ/mol were in the range of reported activation energy data for various foods and have been related to the moisture diffusion. Differences of water mobility within films made of AX compared to films of BG have to be connected with the local composition heterogeneity of cereals cell walls. The endosperm tissue contains a global ratio of AX/BG of 70/30 while other more external and hydrophobic tissues are enhanced in BG content. This observation agrees with a water mobility reduced within BG films, which display more compact mesostructures.

Analyses of the second moment M_2 allowed the investigation, within the polysaccharides films more and less hydrated, of the solid phase mobility. Monitoring of M_2 while temperature and the water content were increased, have emphasized the M_2 decreasing associated with weaker dipolar interactions within the polysaccharide phases. Linear chains of β-Glucans seemed to display a more organised and/or more compact mesostructure than AX within films. Secondly, higher temperatures as well as water contents resulted in the improvement of the anisotropic mobility of the polysaccharides chains which in the same time diffused more rapidly.

Significant impact of the sample preparation (in the form of powder, minced or stacked discs of films) and of their moisture content have been deduced from all these NMR measurements.

This work will continue with the investigation of the impact of the arabinoxylans structure (A/X ratio) on the hydration properties of films made of these polysaccharides. Studies of the water dynamics in composite films containing mixture of AX and 1 BG with varying content should permit to get a better model of the polysaccharide assemblies within cereal cell walls.

Acknowledgement

The authors are indebted to the Regional Council of Pays de la Loire for financial support (PhD grant). They also acknowledge Luc Saulnier (INRA Nantes) for fruitful discussions.

References

1 G.B. Fincher and B.A. Stone, in *Advances in Cereal Science and Technology,* Vol. VIII, ed. Y. Pomeranz, American Association of Cereal Chemists Inc., St Paul.,1986.
2 C.M. Courtin and J.A. Delcour, *Journal of Cereal Science*, 2002, **35**, 225.

3 G. Beresford and B.A. Stone, *Journal of Cereal Science,* 1983, **1**, 111.

4 W. Cui, P.J. Wood, B. Blackwell and J. Nikiforuk, *Carbohydr. Polym.*, 2000, **41**, 249.

5 A.S. Perlin, *Cereal Chemistry*, 1951, **28**, 382.

6 M.S. Izydorczyk and C.G. Biliaderis, *Carbohydr. Polym.*, 1995, **28**, 33.

7 G. Dervilly, L. Saulnier, P. Roger and J. F. Thibault, *J. Agri. Food Chem.*, 2000, **48**,270.

8 G. Dervilly, J. F. Thibault and L. Saulnier, *Carbohydrate Research,* 2001, **330**, 365.

9 M.M. Smith and R.D. Hartley, *Carbohydr. Res.*, 1983, **118**, 65.

10 B. Ahluwalia and S.C. Fry, *Journal of Cereal Science*, 1986, **4**, 287.

11 C.F. Ciacco and B.L. d'Appolonia, *Cereal Chemistry,* 1982, **59**, 163.

12 R.G. Fulcher, S.S. Miller and R. Ruan, in *Functionality of food phytochemicals,* eds. T. Johns and J.T. Romeo, Plenum, New York, 1997, p. 237.

13 C. Barron, M.L. Parker, E.N.C. Mills, X. Rouau and R.H. Wilson, *Planta,* 2005, **220**, 667.

14 S. Philippe, P. Robert, C. Barron, L. Saulnier and F. Guillon *J. Agri. Food Chem.*, 2006, **54**, 2303.

15 S. Philippe, C. Barron, P. Robert, M. F. Devaux, L. Saulnier and F. Guillon, *J. Agri. Food Chem.*, 2006, **54**, 5113.

16 F. Guillon, O. Tranquet, L. Quillien, J.P. Utille, J.J. Ordaz Ortiz and L. Saulnier *Journal of Cereal Science,* 2004, **40**, 167.

17 S. Philippe, L. Saulnier and F. Guillon, *Planta,* 2006, **224**, 449.

18 D. W. Marquardt, *J. Soc. Ind. Appl. Math.,* 1963, **11**, 431.

19 S.W. Provencher: *Comput. Phys. Commun.*, 1982, **27**, 229.

20 A. Abragam, in *The principles of nuclear magnetism*; Clarendon Press: Oxford, 1961.

21 W. Derbyshire, M. Van den Bosch, D. van Dusschoten, W. MacNaughtan, I.A. Farhat, M.A. Hemminga and J.R. Mitchell, *J. Magn. Reson.*, 2004, **168**, 278.

22 I. Van den Dries, D. Van Dusschoten and M. A. Hemminga, *Physical Chemistry B,* 1998, **102**, 10483.

23 R. Partanen, V. Marie, W. Macnaughtan, P. Forssell and I. Farhat *Carbohydrate Polymers*, 2004, **56**(2), 147.

24 G. Roudaut, I. Farhat, F. Poirier-Brulez and D. Champion, *Carbohydrate Polymers*, 2009, **77**, 489.

25 S. Ablett, A. H. Darke, M. J. Izzard and P. J. Lillford, in *The glassy state in foods*, eds. J. M. V. Blanshard and P. J. Lillford, Nottingham: Nottingham Press, 1993, p. 189–206.

26 B. P. Hills and B. Remigereau, *Int. J. Food Sci. and Tech.*,1997, **32**, 51.

27 S. R. Logan, *J. Chem. Educ.*, 1982, **59**, 279.

28 M. Tolaba and C. Suarez, *Lebensmittel-Wissenschaft und-Technologie*, 1988, **21**, 83.

29 J. Bon, S. Simal, C. Rossello and A. Mulet, Journal of Food Engineering, 1997, **34**,109.

30 I. Doymaz, *Journal of Food Engineering*, 2004, **61**, 359.

Nutrition

METABOLIC RESPONSES TO HEAT, ANOXIA, OR OXIDATIVE STRESS ELUCIDATED IN MUSCLE CELL CULTURES USING ^{13}C NMR SPECTROSCOPY

Ida K. Straadt,[1] Jette F. Young,[2] Bent O. Petersen,[3] Jens Ø. Duus,[3] Niels Gregersen,[4] Peter Bross,[4] Niels Oksbjerg,[2] and Hanne C. Bertram[1]

[1]Department of Food Science, Faculty of Agricultural Sciences, University of Aarhus, Kirstinebjergvej 10, DK-5792 Aarslev, Denmark. E-mail: ida.straadt@agrsci.dk
[2]Department of Food Science, Faculty of Agricultural Sciences, University of Aarhus, Blichers Allé 20, P.O. Box 50, DK-8830 Tjele, Denmark
[3]Carlsberg Laboratory, Gamle Carlsberg Vej 10, 2500 DK Valby, Denmark
[4]Research Unit for Molecular Medicine, University of Aarhus, Brendstrupgaardsvej 100, DK-8200, Aarhus N, Denmark

1 INTRODUCTION

Stressful conditions experienced by pigs in relation to slaughter influence the meat quality. The stress comprises the transportation to the abattoir, potential mixing with unfamiliar pigs, lairage and handling at the abattoir, and the slaughter procedure itself.[1-5] The pre-slaughter conditions transportation, mixing, lairage, and handling can lead to increase in the body temperature and thereby result in heat stress.[1,2,6-8] Moreover, stunning and exsanguination cause anoxia and disruption of the energy supply to the muscles, respectively.[7,9] Furthermore, heat and/or anoxic stress experienced in relation to slaughter can also result in oxidative stress.[10] Hence, in relation to slaughter, a range of different stressors are prevailing in the muscle tissue of the pigs.

^{13}C nuclear magnetic resonance (NMR) spectroscopy has not been widely applied in meat science,[11-14] but in pigs exposed to treadmill stress a possible relation between degradation of creatine and formation of lactate was found by application of solid state magic-angle spinning (MAS) ^{13}C NMR spectroscopy in post mortem muscles.[14] ^{13}C NMR spectroscopy has been applied in research on cancer cells,[15-19] and also in relation to stress research on cancer cells.[16,17] Since the natural abundance of the ^{13}C nucleus is low,[20] it is usually necessary to incubate with ^{13}C isotopes to get sufficient signals. By applying a muscle cell line as a model for muscles, different stressors of interest in relation to slaughter can be investigated.[21-23] By exposing the muscle cells to different stressors while incubating the cells with a ^{13}C isotope, it is possible to asses the metabolic changes in response to these stressors by application of ^{13}C NMR spectroscopy.[21,22]

2 METHOD AND RESULTS

2.1 Experimental setup and ^{13}C NMR spectroscopy

In the present study multinuclear myotubes from the mouse cell line C2C12 were exposed to heat or anoxic stress (Figure 1) or to oxidative stress (Figure 2), to imitate stressful conditions apparent in relation to slaughter. By incubating the myotubes with [^{13}C$_1$]glucose for 4 h, the metabolism of glucose can be monitored by application of ^{13}C NMR spectroscopy. The myotubes were extracted with a methanol/chloroform/water extraction method adopted from Serkova et al.[24]

The cell extracts from two Petri dishes (the samples depicted in Figure 1) or from one Petri dish (the samples depicted in Figure 2) were dissolved in 600 μL D_2O containing 0.025% (w/v) sodium trimethylsilyl-[2,2,3,3-D_4]-1-propionate (TSP) and 0.01% (v/v) dioxane. ^{13}C NMR spectra of the extractions of the myotubes were recorded at 25 °C on a Bruker Avance 800 spectrometer, operating at a frequency of 201.01 MHz and equipped with a 5-mm ^{1}H observe cryoprobe (Bruker BioSpin, Rheinstetten, Germany). Proton-decoupled ^{13}C NMR spectra were recorded using Bruker standard homo-nuclear pulse programs, and each spectrum was the sum of 1024 free induction decays (FID). A 90° pulse was applied with a repetition time of 3 s and an acquisition time of 0.682 s. All spectra were referenced to TSP at 0 ppm. Dioxane (δ 69.44 ppm) was used as a secondary reference in the ^{13}C NMR spectra. ^{13}C NMR spectra were obtained on three replicates of each of the treatment groups described in Figure 1, and on an additional sample for the control group. Likewise, ^{13}C NMR spectra were obtained on three replicates of each treatment groups described in Figure 2.

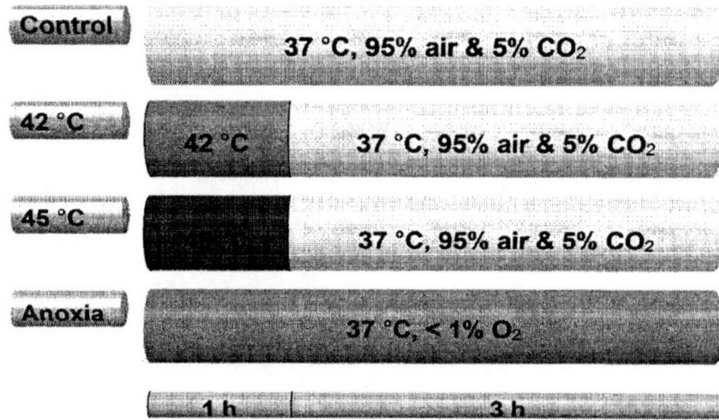

Figure 1 *Experimental setup for the C2C12 mouse myotubes exposed to heat or anoxic stress. The myotubes were incubated with [$^{13}C_1$] glucose for 4 h, and during this incubation period exposed to the different treatments.*

Figure 2 *Experimental setup for the C2C12 mouse myotubes exposed to oxidative stress. The myotubes were incubated with [$^{13}C_1$] glucose for 4 h, and during this incubation period exposed to the different treatments.*

Quantification of the various metabolites was carried out by integration of peak areas using the Topspin software (Bruker BioSpin, Rheinstetten, Germany). The ^{13}C NMR spectra were integrated in the spectral range 19.00-100.00 ppm, excluding the dioxane signal at 69.00-70.00 ppm. All the data were analyzed by using the mixed procedure of SAS (SAS Institute Inc., Cary, NC, USA). The model included treatment as a fixed effect and replicates as a random effect. Absolute integral intensities are given as least squares means (LSMeans).

2.2 Metabolic changes in myotubes exposed to heat, anoxia or oxidative stress

In all the ^{13}C NMR spectra the ^{13}C label was found in the metabolites $[^{13}C_3]$alanine, $[^{13}C_2]$, $[^{13}C_3]$, and $[^{13}C_4]$glutamate, $[^{13}C_2]$ and $[^{13}C_3]$aspartate, $[^{13}C_3]$lactate, $[^{13}C_2]$ and $[^{13}C_4]$glutamine, $[^{13}C_2]$ and $[^{13}C_4]\gamma$-glutamyl-glutathione and $[^{13}C_1]$fructose-6-phosphate (F-6-P) (Figure 3 and Table 1). In addition, taurine signals from naturally abundant ^{13}C were detected in all the ^{13}C NMR spectra in C_1 and C_2 taurine (Figure 3 and Table 1). At anoxic conditions the ^{13}C label was also found in the metabolites $[^{13}C_1]$ and $[^{13}C_3]$glycerol-3-phosphate (Gly-3-P), and $[^{13}C_1]$ and $[^{13}C_6]$fructose-1,6-bisphosphate (F-1,6-P) (Figure 1B and Table 1). All the metabolites identified in the ^{13}C NMR spectra were metabolites derived from the glycolysis or the Krebs cycle (Figure 4), except for taurine.

Exposure to either heat, anoxic, or oxidative stress resulted in lower levels of $[^{13}C_2]$, $[^{13}C_3]$, and $[^{13}C_4]$glutamate, $[^{13}C_2]$ and $[^{13}C_4]$glutamine, $[^{13}C_2]$ and $[^{13}C_4]\gamma$-glutamyl-glutathione, $[^{13}C_2]$ and $[^{13}C_3]$aspartate, and $[^{13}C_1]$F-6-P (Figure 3 and Table 1). Decrease in natural abundance ^{13}C taurine was also observed. The decreases in these metabolites were generally small after exposure to 42 °C or 100 µM H_2O_2, more apparent after 45 °C or 500 µM H_2O_2 exposures, and most pronounced after anoxic stress, when compared to standard incubation conditions (Table 1). Hence, for the stressors explored in the present experiment heat and oxidative stress caused some impairment of the glucose metabolism, whereas anoxia had the most severe impact on the glucose metabolism. These results are in agreement with effects of stress observed in glioma and hepatoma cells exposed to anoxia.[16]

Conversely, higher levels of $[^{13}C_3]$lactate were observed after heat, oxidative, or anoxic stress exposure (Figure 3 and Table 1), which was detectable after heat and oxidative stress, but more pronounced after anoxic stress. This indicates that anaerobe metabolism increases at heat and oxidative stress, and that the increase is most pronounced after anoxic stress exposure. Anoxia also resulted in accumulation of the phosphomonoesters $[^{13}C_1]$ and $[^{13}C_3]$Gly-3-P, and $[^{13}C_1]$ and $[^{13}C_6]$F-1,6-P (Figure 3B and Table 1). The Gly-3-P functions together with dihydroxyacetone phosphate as a shuttle to transport electrons from the cytosol to the mitochondria (see Figure 4). Hence, the appearance of Gly-3-P in the ^{13}C NMR spectra at anoxic stress is consistent with a blockage of this shuttle and a shift to anaerobic metabolism due to inhibition of the aerobic pathway in the mitochondria, which in turn leads to accumulation of F-1,6-P (see Figure 4). Appearance of phosphomonoesters at anoxic stress is in agreement with findings in rat hearts, where an increase in Gly-3-P was found with increasing exposure time to anoxic conditions.[25]

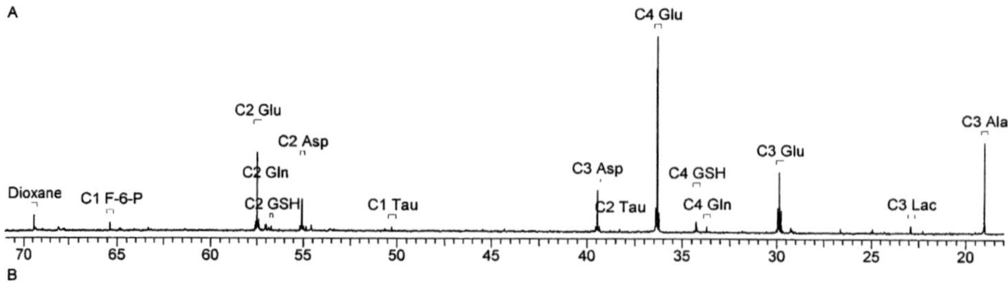

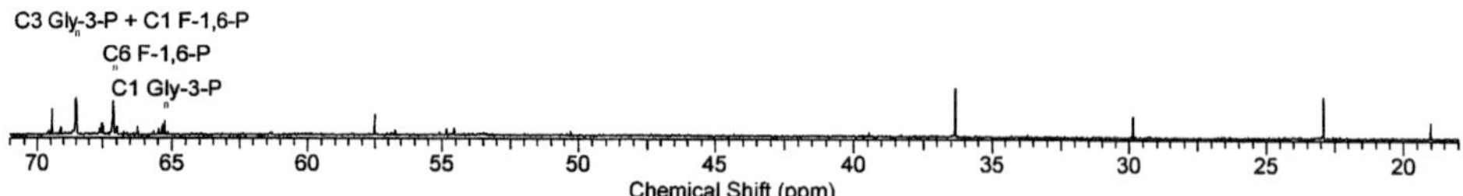

Figure 3 *Representative ^{13}C NMR spectra of the C2C12 mouse myotubes after incubation with $[^{13}C_1]$ glucose for 4 h. The spectra were obtained on the freeze-dried methanol/water extractions of control myotubes (A) or myotubes exposed to anoxic stress (B). F-6-P: fructose-6-phosphate; Glu: glutamate; Gln: glutamine; GSH: γ-glutamyl-glutathione; Asp: aspartate; Tau: taurine; Lac: lactate; Ala: alanine; Gly-3-P: glycerol-3-phosphate; F-1,6-P: fructose-1,6-bisphosphate. Dioxane is an internal reference.*

Heat or anoxia caused a decrease in $[^{13}C_3]$alanine, whereas oxidative stress resulted in an increase in $[^{13}C_3]$alanine (Table 1). Hence, the present study also shows that different stressors result in different alternations in metabolites.

Anoxia resulted in the most severe changes in the metabolites, which was apparent as the largest decreases in metabolites, mainly amino acids, and accumulation of Gly-3-P (Figure 3B and Table 1). This is not surprising since the anoxic stress exposure was for 4 h, while 1 h stress exposure was applied for the other stressors (Figure 1 and 2).

Furthermore, anoxia also results in anaerobic conditions, which obviously influence the glucose metabolism markedly.

2.3 Metabolites as markers for stressors of interest in relation to slaughter stress

The decreases in the metabolites glutamate, glutamine, γ-glutamyl-glutathione, aspartate, F-6-P, and taurine observed at all the different stress exposures could possibly be used as markers for stress conditions of interest in relation to slaughter.[22] These findings are in agreement with findings in pig hearts exposed to ischemia/reperfusion injury.[26]

Conversely, increase in lactate could be utilized as a marker for stress, and accumulation of Glyl-3-P and F-1,6-P are potential markers distinct for anoxic stress. In other studies Gly-3-P has been found to be one of the most prominent markers for ischemia/reperfusion recovery in rat hearts.[25]

Alanine decreased after exposure to heat or anoxic stress, but increased after oxidative stress exposure. Hence, the present studies also indicate that different stressors result in different alternations in metabolites, and one needs to be careful in choice of metabolic markers for different stressors.

Table 1.   *^{13}C NMR chemical shifts and LSMeans of the absolute integral intensities. Metabolite abbreviations as in Figure 3*

Metabolite / Shift (ppm)	Control Heat / anoxia (n=4)	Heat 42 °C (n=3)	Heat 45 °C (n=3)	Anoxia < 1% O_2 (n=3)	Control H_2O_2 (n=3)	H_2O_2 100 μM (n=3)	H_2O_2 500 μM (n=3)
Ala (19.04)	3.71	3.62	1.69[a]	0.49[a]	3.21	3.65	5.15[a]
Lac (22.93)	0.25	1.20	0.67	2.03[a]	0.37	0.56	0.69
Glu (29.89)	7.10	5.24[b]	2.15[b]	0.57[b]	14.27	13.89	8.74[a]
Gln (33.73)	0.26	0.19	0.07[a]	0.03[a]	0.38	0.23	0.06[a]
GSH (34.28)	1.01	0.86	0.41[a]	0.02[b]	1.50	1.22	1.00
Glu (36.36)	13.75	9.47[a]	4.05[b]	1.60[b]	26.87	26.07	15.31[a]
Tau (38.29)	0.18	0.17	0.05[a]	0.06[a]	0.12	0.07	0.02
Asp (39.48)	3.43	2.12[a]	0.74[b]	0.09[b]	5.20	5.80	4.23
Tau (50.27)	0.31	0.18	0.05[a]	0.13[a]	0.56	0.34	0.01
Asp (55.08)	2.89	1.59[a]	0.66[a]	0.06[b]	4.39	4.49	2.42
GSH (56.85)	0.39	0.29	0.21	0.10	0.82	0.63	0.01
Gln (57.10)	1.01	0.63	0.27[a]	0.05[a]	1.39	0.95	0.01
Glu (57.53)	6.93	4.70[a]	2.05[b]	0.53[b]	13.14	12.37	5.74[a]
Gly-3-P (65.27)	Nd[c]	Nd[c]	Nd[c]	0.64[a]	Nd[c]	Nd[c]	Nd[c]
F-6-P (65.37)	0.46	0.40	0.14[a]	0.10[a]	0.55	0.55	0.12
F-1,6-P (67.12)	Nd[c]	Nd[c]	Nd[c]	2.14[a]	Nd[c]	Nd[c]	Nd[c]
Gly-3-P / F-1,6-P (68.51)	Nd[c]	Nd[c]	Nd[c]	2.53[a]	Nd[c]	Nd[c]	Nd[c]

[a] LSMeans differ significantly ($P < 0.05$) when compared with the heat/anoxia or the H_2O_2 control cells, respectively.
[b] LSMeans differ significantly ($P \leq 0.0001$) when compared with the heat/anoxia control cells.
[c] Not detectable.

3 CONCLUSION

By applying a muscle cell model one stressor at a time can be investigated under controlled conditions. This gives an impression of which metabolites that alternates with different stressors, and which metabolites that could be potential markers for stress of interest in relation to slaughter. Hence, from the present study it is apparent that decreases in the metabolites glutamate, glutamine, γ-glutamyl-glutathione, aspartate, fructose-6-phosphate, and taurine, and conversely increase in lactate and accumulation of glycerol-3-phosphate and fructose-1,6-bisphosphate could be potential markers for slaughter stress.

Acknowledgement

The Danish Technology and Production Research Council (FTP) is gratefully acknowledged for financial support through the projects "Cellular stress and metabolic responses to ante- and post mortem stress factors elucidated in primary porcine muscle cell cultures using confocal microscopy and NMR-based metabolic profiling" (project no. 274-06-0107) and NMR-based metabonomics on tissues and biofluids (project no. 274-05-339).

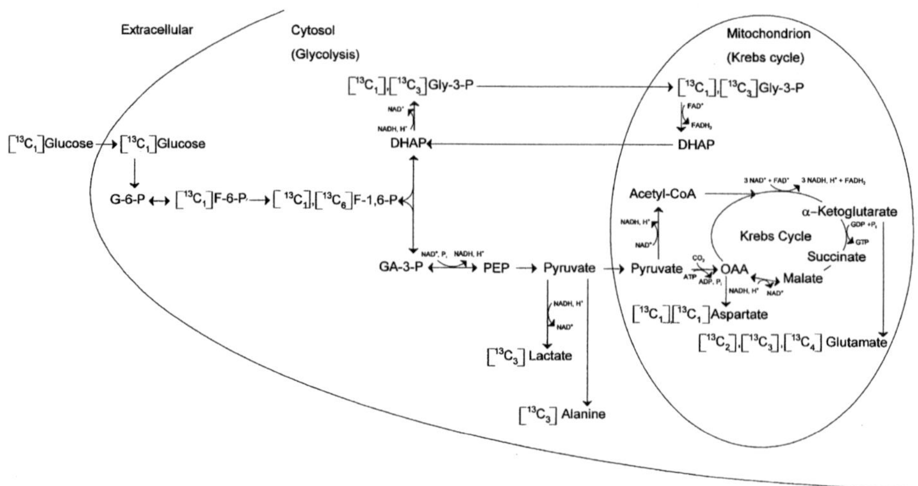

Figure 4 *The main reactions involved in [¹³C₁]glucose metabolism in the C2C12 mouse myotubes are depicted. The aerobic pathway occurs when pyruvate is oxidized in the Krebs cycle. At heat, anoxia, and oxidative stress, an increase in [¹³C₁]lactate production is occurring through the anaerobic glycolytic pathway, and anoxia also leads to [¹³C₁],[¹³C₃]Gly-3-P and [¹³C₁],[¹³C₆] F-1,6-P accumulation. Gly-3-P: glycerol-3-phosphate; DHAP: dihydroxyacetone phosphate; G-6-P: glucose-6-phosphate; F-6-P: fructose-6-phosphate; F-1,6-P: fructose-1,6-bisphosphate; GA-3-P: glyceraldehyde-3-phosphate; PEP: phosphoenolpyruvate; OAA: oxaloacetate.*

References

1   P. G. van der Wal, B. Engel and H. G. M. Reimert, *Meat Sci.*, 1999, **53**, 101.
2   S. Stoier, M. D. Aaslyng, E. V. Olsen and P. Henckel, *Meat Sci.*, 2001, **59**, 127.
3   D. Alvarez, M. D. Garrido and S. Banon, *Food Res. Int.*, 2009, **25**, 233.
4   J. M. Yu, S. Tang, E. D. Bao, M. Zhang, Q. Q. Hao and Z. H. Yue, *Meat Sci.*, 2009, **83**, 474.
5   Q. W. Shen, W. J. Means, S. A. Thompson, K. R. Underwood, M. J. Zhu, R. J. McCormick, S. P. Ford and M. Du, *Meat Sci.*, 2006, **74**, 388.
6   J. F. Young, H. C. Bertram and N. Oksbjerg, *Meat Sci.*, 2009, **83**, 634.
7   C. Terlouw, *Livest. Prod. Sci.*, 2005, **94**, 125.
8   U. Kuchenmeister, G. Kuhn and K. Ender, *Meat Sci.*, 2005, **71**, 690.
9   T. Grandin, *Int. J. Stud. Anim. Prob.*, 1980, **1**, 313.
10 B. Halliwell, *Free Radic. Biol. Med.*, 2009, **46**, 531.
11 P. Lundberg, H. J. Vogel and H. Ruderus, *Meat Sci.*, 1986, **18**, 133.
12 H. C. Bertram, H. J. Jakobsen, H. J. Andersen, A. H. Karlsson and S. B. Engelsen, *J. Agric. Food Chem.*, 2003, **51**, 2064.

13 H. C. Bertram and H. J. Andersen, in *Annual Reports on NMR Spectroscopy*, 53rd Edn., ed. G. A. Webb, Academic Press, 2004, p. 157.

14 H. C. Bertram, H. J. Jakobsen and H. J. Andersen, *J. Agric. Food Chem.*, 2004, **52**, 3159.

15 S. Gottschalk, N. Anderson, C. Hainz, S. G. Eckhardt and N. J. Serkova, *Clin. Cancer Res.*, 2004, **10**, 6661.

16 A. Perrin, E. Roudier, H. Duborjal, C. Bachelet, C. Riva-Lavieille, X. Leverve and R. Massarelli, *Biochimie*, 2002, **84**, 1003.

17 A. Perrin, H. Duborjal, E. Roudier and R. Massarelli, *Arch. Ital. Biol.*, 2003, **141**, 1.

18 A. K. Bouzier, P. Voisin, R. Goodwin, P. Canioni and M. Merle, *Dev. Neurosci.*, 1998, **20**, 331.

19 A. L. Merz and N. J. Serkova, *Biomark. Med.*, 2009, **3**, 289.

20 H. J. Issaq, Q. N. Van, T. J. Waybright, G. M. Muschik and T. D. Veenstra, *J. Sep. Sci.*, 2009, **32**, 2183.

21 I. K. Straadt, J. F. Young, B. O. Petersen, J. O. Duus, N. Gregersen, P. Bross, N. Oksbjerg and H. C. Bertram, *Metabolism*, 2010, **59**, 814.

22 I. K. Straadt, J. F. Young, B. O. Petersen, J. O. Duus, N. Gregersen, P. Bross, N. Oksbjerg, P. K. Theil and H. C. Bertram, *J. Agric. Food Chem.*, 2010, **58**, 1918.

23 I. K. Straadt, J. F. Young, P. Bross, N. Gregersen, N. Oksbjerg, P. K. Theil and H. C. Bertram, *J. Agric. Food Chem.*, 2010, **58**, 6376.

24 N. Serkova, J. Klawitter and C. U. Niemann, *Transpl. Int.*, 2003, **16**, 748.

25 F. M. H. Jeffrey, C. J. Storey and C. R. Malloy, *Basic Res. Cardiol.*, 1992, **87**, 548.

26 I. Barba, E. Jaimez-Auguets, A. Rodriguez-Sinovas and D. Garcia-Dorado, *Magn. Reson. Mater. Phy.*, 2007, **20**, 265.

ASSESSMENT OF THE PERFORMANCE OF AN NMR ASSAY FOR THE DETERMINATION OF HYDROPEROXIDES IN LIPIDS

Christina Skiera[1], Panagiotis Steliopoulos[1], Ulrike Holzgrabe[2], Bernd Diehl[3]

[1] Chemisches und Veterinäruntersuchungsamt Karlsruhe, Weißenburger Str. 3, 76187 Karlsruhe, Germany
[2] Institute of Pharmacy and Food Chemistry, University of Würzburg, Am Hubland, Würzburg Germany
[3] Spectral Service, Emil-Hoffmann-Str. 33, 50996 Köln, Germany

1 INTRODUCTION

Hydroperoxides that are formed as primary products during the autoxidation process of oils and fats are analytical markers for lipid deterioration. ^{1}H-NMR spectroscopy enables to detect and quantify all hydroperoxide species in oils and fats simultaneously. Generally, the level of primary oxidation products is indicated by the peroxide value (PV) according to Wheeler. A comparison between the analytical performance of the Wheeler method and the NMR approach seems to be intractable. The results obtained by the Wheeler method are affected by an inherent systematic error which is ignored per definition. Additionally, the precision of the PV is expressed in *meq/kg* and the precision of the hydroperoxide amount by NMR spectroscopy can be given in *mmol/kg*, which makes a comparison rather difficult. In this contribution, a statistical procedure is presented that copes with this difficulties.

2 MATERIAL AND METHODS

Peroxide-containing oil: Methyllinoleate (purchased from Sigma) was oxidized at 70°C in the dark in an open vessel.

Peroxide-free oil: Delios® V (supplied by Cognis, Mannheim) is an artificial triacylglycerol consisting mainly of caprylic acid and capric acid.

Mixtures of the peroxide-free and the peroxide-containing oil were prepared with the following mixing ratios: 1:26, 1:16, 1:10 and 1:7.

The PV was determined according to the international standard ISO 3960:2007.[1]

^{1}H-NMR: The mixtures of the lipids were dissolved in $CDCl_3$/DMSO-d_6 (5:1 v/v) and placed in a 5 mm diameter tube. ^{1}H-NMR spectra were recorded on a Bruker 400 MHz spectrometer operating at 400.17 MHz. Acquisition parameters: spectral width 20.5396 ppm, relaxation delay 1 s, acquisition time 7.97 s, number of scans 128. The experiment was carried out at 300 K using a flip angle of 30°. The hydroperoxide content was calculated from the hydroperoxide proton integrals at δ = 10 to 11 ppm in the spectrum.

3 RESULTS AND DISCUSSION

In Fig. 1 the PVs are plotted versus the NMR-determined peroxide amounts and a straight line is fitted to the data. Taking into account that both quantities are subject to error, Deming regression was applied. The functional relationship between the PV and the NMR-determined peroxide amount was found to be given by the equation:

$$y = 2.42x + 1.10$$

To compare the methods, the relative sensitivity[2] of the NMR assay with respect to the Wheeler method, RS _NMR/Wheeler_, was calculated. This ratio expresses which method has a greater ability to detect a difference in the actual hydroperoxide amount. The obtained result RS _NMR/Wheeler_ = 0.9 clearly indicates that both methods exhibit a similar analytical performance.[3]

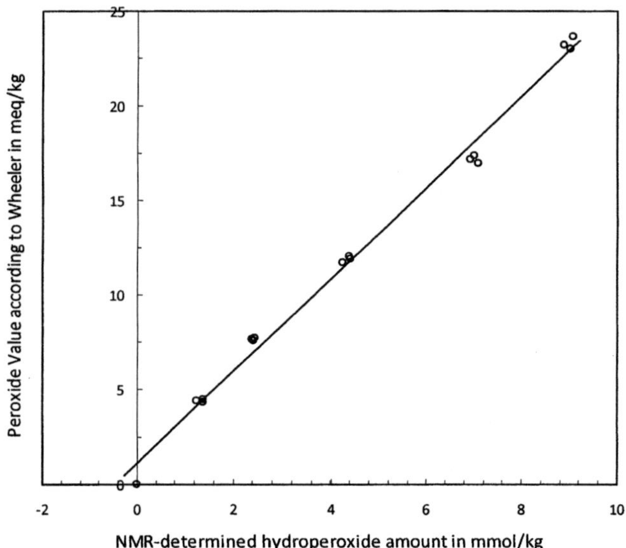

Figure 1   *PV in meq/kg versus the NMR-determined peroxide amount in mmol/kg (curve fitting by Deming regression)*

References

1   ISO 3960:2007 Animal and vegetable fats and oils - *Determination of peroxide value - Iodometric (visual) endpoint determination.*
2   J. Mandel, *The Statistical Analysis of Experimental Data.* Courier Dover Publications, New York, 1984, ch. 14, p. 363.
3   C. Skiera, P. Steliopoulos, T. Kuballa, B. Diehl, U. Holzgrabe, publication in preparation

APPLICATION OF ELECTRON SPIN RESONANCE SPECTROSCOPY TO STUDY DIETARY INGREDIENTS AND SUPPLEMENTS - DUAL ANTIOXIDANT AND PROOXIDANT FUNCTIONS OF RETINYL PALMITATE

J.J. Yin,[1,*] Q. Xia,[2] H. Lutterodt,[3] W. Wamer,[1] and P.P. Fu[2]

[1]Center for Food Safety and Applied Nutrition, U.S. Food and Drug Administration, College Park, MD 20740, USA
[2] National Center for Toxicological Research, U.S. Food and Drug Administration, Jefferson, AR 72079, USA
[3] Department of Nutrition and Food Science, University of Maryland, College Park, MD 20742, USA

*Corresponding author. Tel: 301-436-1991, E-mail: junjie.yin@fda.hhs.gov

This article is not an official guidance or policy statement of U.S. Food and Drug Administration (FDA). No official support or endorsement by the U.S. FDA is intended or should be inferred.

1 INTRODUCTION

It is well established that many components of our diets exhibit antioxidant and/or pro-oxidant activities.[1] In most cases, the antioxidant and pro-oxidant functions of food substances are associated with quenching or enhancing the formation of reactive oxygen species (ROS). This function is important since there is an association between elevated ROS and age-associated diseases.[2] Therefore, it is timely and important to determine antioxidant and pro-oxidant activities of foods and their chemical constituents, and elucidate their mechanisms of action.

Electron spin resonance (ESR) spectroscopy is an important technique for directly detecting free radicals generated chemically and biologically. ESR spin trapping and ESR spin labelling are the methodologies most commonly used for detection of free radical formation and their identification.[1] We have a longstanding interest in employing ESR techniques to quantify free radical formation in biological systems and to study the mechanisms by which dietary constituents exhibit anti- and/or pro-oxidant activities.[3-12]

Vitamin A is an essential human nutrient that plays an important role in many physiological functions. Retinyl palmitate (RP) is one of the primary storage forms of vitamin A.[13,14] In this review paper, we use ESR techniques to provide evidence that RP can act either as an antioxidant agent that scavenges free radicals, or upon UV irradiation, as a pro-oxidant agent that generates free radicals. The ESR techniques discussed in this paper include: (i) ESR spin trapping - utilizing specific spin trapping agents to react with short-lived free radicals, resulting in the formation of adducts that have half-lives long enough for ESR spectroscopic measurement,[3,4,8-12,15-17] and (ii) ESR spin labelling- using a stable paramagnetic spin label agent to interact with the target chemical to probe the structural and/or dynamic changes in a specific chemical/biological system.[5,7,8]

2 RETINYL PALMITATE ACTS AS AN ANTIOXIDATANT

2.1 Scavenging of Hydroxyl Radicals

Hydroxyl radical (HO$^{\bullet}$) was generated by the Fenton reaction, involving the reaction of 0.2 mM FeSO$_4$ and 0.2 mM H$_2$O$_2$. The spin trap 5,5-dimethyl *N*-oxide pyrroline (DMPO, 50 mM) was added to the reaction mixture. Reaction of DMPO with hydroxyl radicals formed the DMPO-OH adduct which exhibited a characteristic four-line (1:2:2:1) ESR spectral profile, with hyperfine splitting parameters $a^N = a^H = 14.9$ G (Figure 1). Upon treatment with RP (1-4 mg/ml), the resulting DMPO-OH ESR signal intensity was reduced. The intensity of the ESR signal decreased with increasing concentration of RP. These results indicate that hydroxyl radicals are efficiently scavenged by RP (Figure 1).

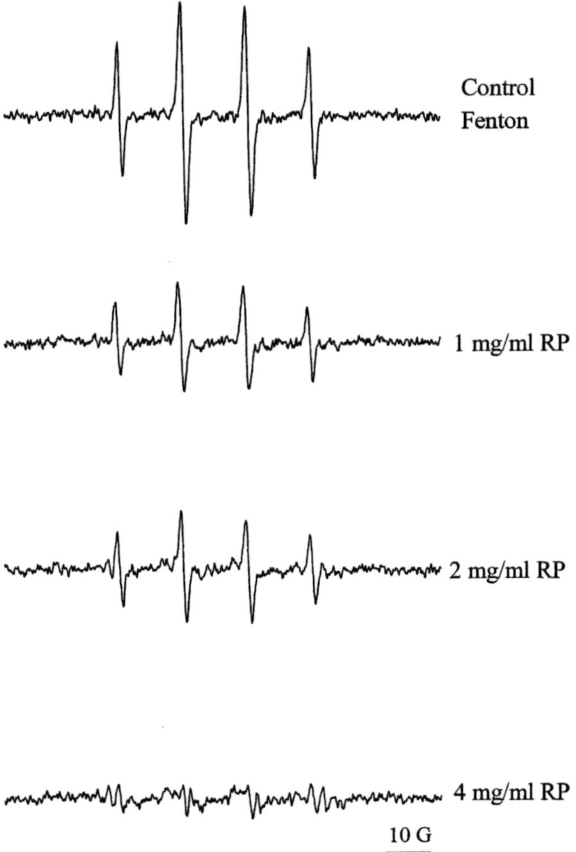

Figure 1 *Scavenging of hydroxyl radicals by RP.*

2.2 Scavenging of Superoxide Radical Anions

A superoxide radical anion generating system containing 1 mM xanthine and 0.05 U/mL xanthine oxidase in 20% acetonitrile in water, 25 mM spin trap 5-*tert*-butoxycarbonyl 5-methyl-1-pyrroline *N*-oxide (BMPO), 0.1 mM diethylenetriaminepentaacetic acid (DTPA) was used. RP (0, 1, 2, and 4 mg/ml, respectively) was added and the mixture was incubated at ambient temperature for 10 min. Figure 2 shows that reaction of BMPO with superoxide radical anion generated the ESR spectral profile of the highly stable BMPO-superoxide adduct (i.e., BMPO-˙OOH). The intensity of the ESR signal decreased in proportion to the amount of RP added indicating that RP scavenges superoxide radical anion in a dose dependent manner.

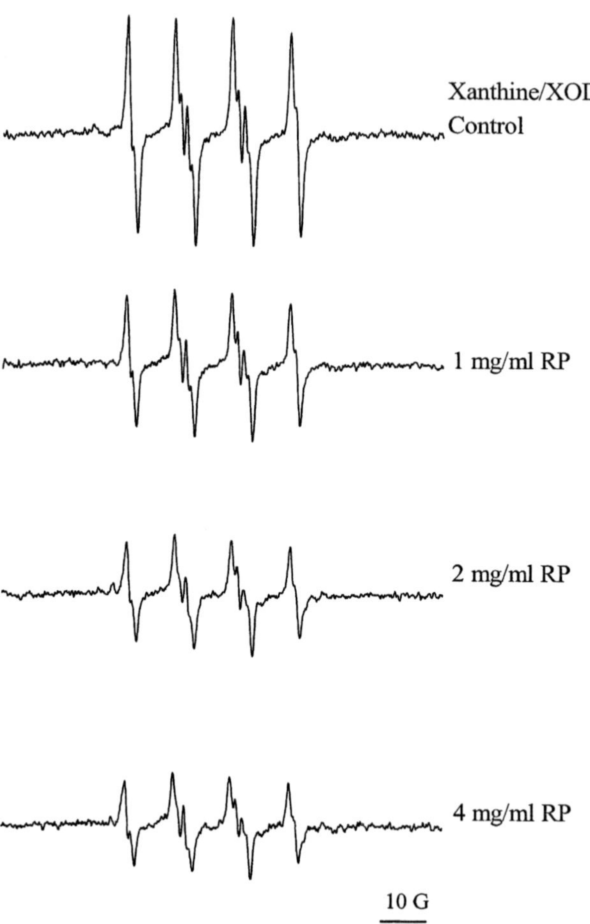

Figure 2 *Scavenging of superoxide radical anions by RP.*

2.3 Scavenging of the Nitrogen-centered Free Radical DPPH by RP

2,2'-Bipyridyl, 2,2-diphenyl-1-picryhydrazyl radical (DPPH˙) is a stable, nitrogen-centered radical that has no involvement in physiological processes. However, attenuation of the ESR signal for DPPH˙ is one of the methods widely used to demonstrate a chemical's ability to scavenge free radicals through donation of a hydrogen atom or, in some cases, electron transfer to DPPH˙. The ESR spectra of DPPH˙, measured without and with RP, are shown in Figure 3. Under our experimental conditions, DPPH˙ itself clearly revealed a strong ESR signal characteristic of DPPH•. Mixture of 0.1 mM DPPH in 20% ethanol with RP (1.25, 2.5, and 5.0 mg/ml) effectively reduced the signal intensity of DPPH•. The reduction in ESR signal intensity was dependent on the concentration of RP.

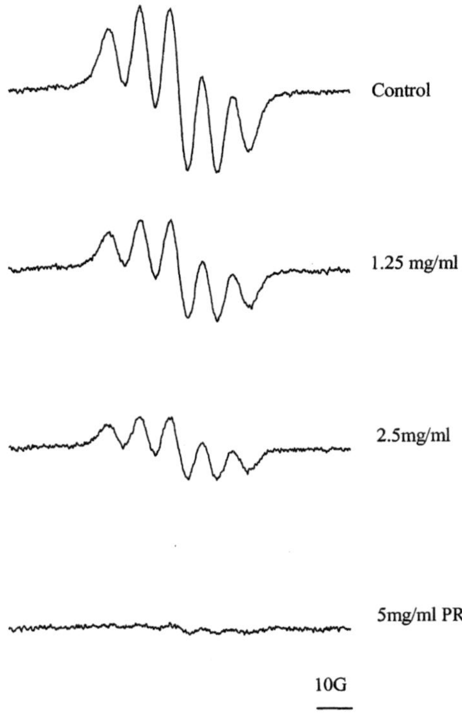

Figure 3 *Scavenging of nitrogen-centered free radical DPPH• by RP*

2.4 Inhibition of Lipid Peroxidation Measured by ESR Oximetry

The ESR oximetry measurement is based on the bimolecular collision of molecular oxygen with a spin probe (spin label) which is a stable free radical. Collision of the spin probe with O_2 produces a spin exchange, resulting in shorter relaxation times and ESR signals with broader line widths and decreasing peak height. Thus, oxygen consumption, such as by lipid peroxidation, results in decreased line widths (and increased peak height) for the spin probe, here 4-oxo-2,2,6,6-tetramethylpiperidine-d_{16}-1-oxyl (^{15}N-PDT). Repeated measurement of the spin probe's line widths allows us to determine rate of lipid peroxidation in the sample. As shown in Figure 4, the progressive increases in peak to peak signal intensity (and accompanying progressive narrowing of line width) in each panel are due to time-dependant reduction of oxygen concentration caused by lipid peroxidation. Addition of RP results in broader line widths, and accompanying lower peak intensities indicating lower consumption rates of O_2 due to inhibition of lipid peroxidation (Figure 4, center and right panels). These results clearly illustrate that RP can inhibit lipid peroxidation induced by 2, 2'-azobis(2,4-dimethylvaleronitrile) (AMVN) in a dose dependent manner.

   Magnetic Resonance in Food Science

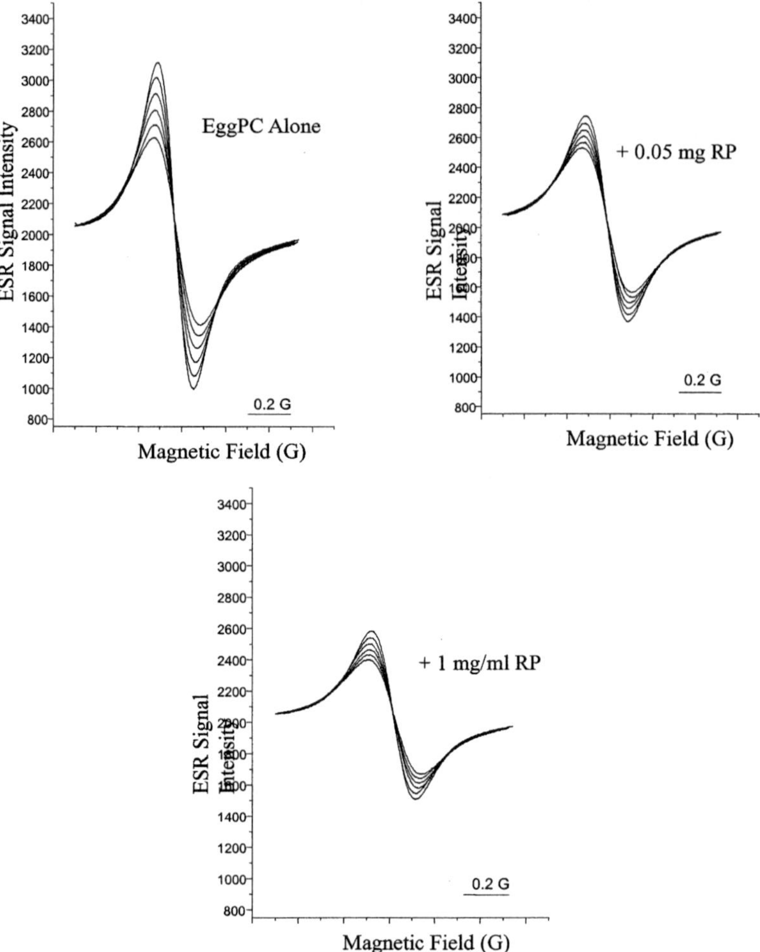

Figure 4   *Measurement of lipid peroxidation by ESR oximetry in a closed chamber. The results are time dependent ESR signals of lipid peroxidation generated in pure EggPC or EggPC mixed with 0.05 and 0.1 mg RP liposomes. ESR spectra, representing the low field line of the spin label, ^{15}N-PDT, were recorded every three min. ESR signals were recorded with 0.5 mW incident microwave and 100 kHz field modulation of 0.05G at 37 °C. Lipid peroxidation was initiated with AMVN.*

3 RETINYL PALMITATE ACTS AS A PRO-OXIDANT

We have been interested in studying the mechanisms through which RP and UVA (320 – 400 nm) light may induce phototoxicity.[14,18-22] Our mechanistic studies on photoexcitation of RP with UVA light demonstrated the formation of 14 photodecomposition products, including anhydroretinol (AR) and 5,6-epoxy-RP (Figure 5).[19] Photoexcitation of RP and these two photodecomposition products, AR and 5,6-epoxy-RP, with UVA in the presence of a lipid, methyl linoleate, resulted in lipid peroxidation which was dependent on the dose of UVA light.[19,20,21,22] In addition, we employed ESR spin-trap techniques to unambiguously confirm that photoexcitation of RP with UVA light produces ROS, which mediated the formation of lipid peroxides.[19,20,21] The ROS identified included singlet oxygen and superoxide radical anion. Similar studies on 5,6-epoxy-RP and AR indicated that these products of RP's photodecomposition exhibited photosensitizing activities similar to those of RP.[19,21,22] For illustration, the ESR experimental results showing the production of superoxide radical anion following photoexcitation of RP with UVA light are shown in Section 3.1.

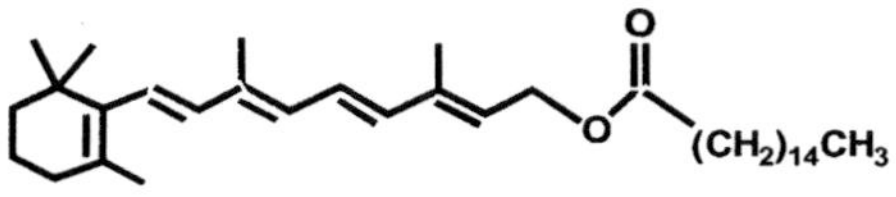

Retinyl palmitate

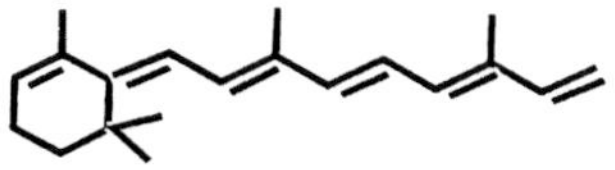

Anhydroretinol

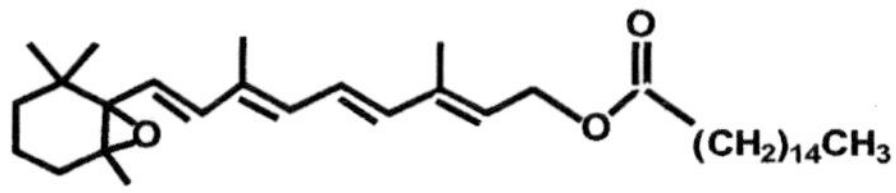

5,6-Epoxyretinyl palmitate

Figure 5 *Chemical structures of RP, anhydroretinol (AR) and 5,6-epoxy-RP.*

3.1 Generation of Superoxide Radical Anion from Photoexcitation of RP

Mixtures containing 0.435, 0.875, 1.750, or 3.500 mg/ml RP in 70% ethanol and 25 mM BMPO at ambient temperature were exposed to UVA light for time intervals up to 1600 sec. The results indicated that the level of superoxide radical anion formed is dependent on both irradiation time and the concentration of RP.

3.2 Formation of Singlet Oxygen from Photoexcitation of RP

The formation of singlet oxygen was studied using the spin trap, TEMP. Solutions containing 4 mg/ml RP in 95% CH_3CN and 5 mM TEMP were irradiated with UVA light at ambient temperature for time intervals up to 60 min. The results indicated that the amount of singlet oxygen generated is dependent on the length of time that samples were exposed to UVA light.[21,22]

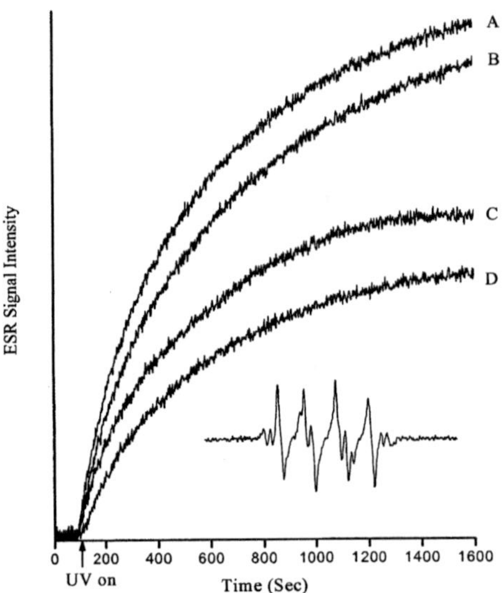

Figure 6 *Real time experiments on superoxide radical generation using different concentrations of RP exposed to UVA light (320 nm). A, 3.5; B, 1.75; C, 0.875 and D, 0.435 mg/ml RP in 70% ethanol and 25 mM BMPO. The insert is the BMPO-O_2^- spin adduct.*

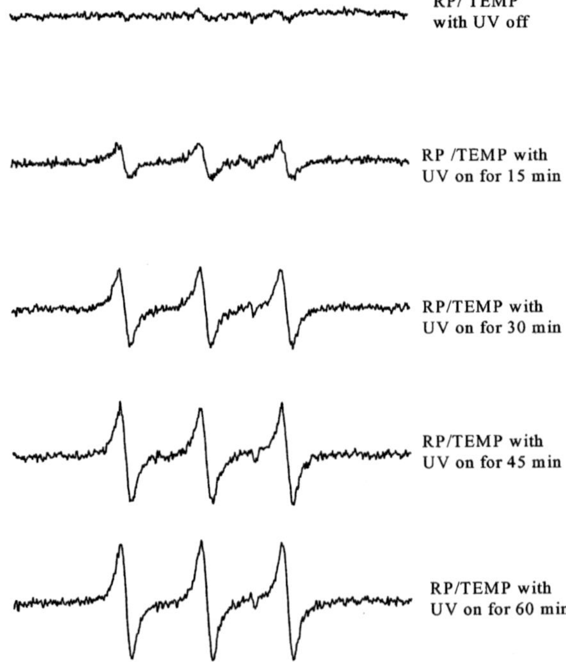

Figure 7 *The time-dependent formation of singlet oxygen by RP irradiated with UVA light (320 nm). During irradiation with UVA light, the ESR signal was recorded every 15 min for up to 60 min. The sample contained 4 mg/ml RP.*

3.3 Generation of Lipid Peroxides during Photoexcitation of RP with UVA Light

ESR oximetry (see Section 2.4) was used to demonstrate that photoexcitation of RP with UVA light induces lipid peroxidation. Lipid peroxidation was measured as a increase in intensity and concomitant decrease in line width of the spin label ^{15}N-PDT. As shown in Figure 8, the progressive increases in peak to peak signal intensity (and accompanying progressive narrowing of line width) in each panel are due to time-dependant increase in oxygen consumption caused by lipid peroxidation. Addition of RP (center and right hand panels) results in higher levels of O_2 consumption due to increased lipid peroxidation induced by photoexcited RP (seen most clearly in the right hand panel as higher peak intensity). These results illustrate that photoexcitation of RP with UVA light can induce lipid peroxidation in a dose dependent manner.

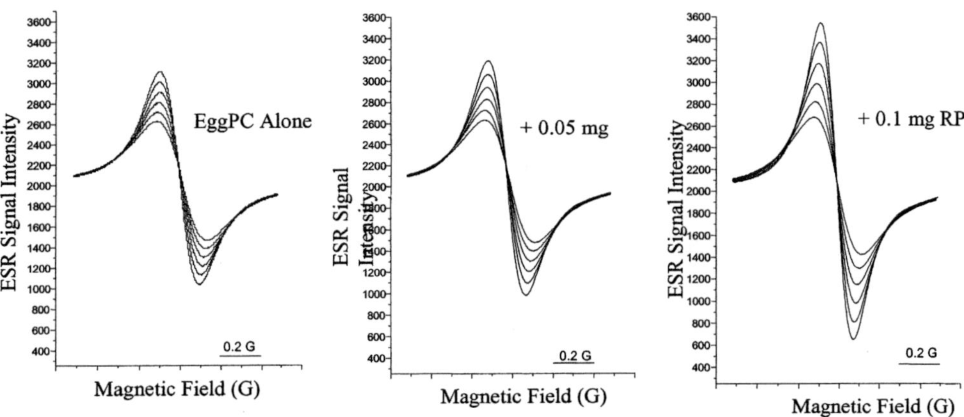

Figure 8 *Time dependence of ESR signal during lipid peroxidation using pure EggPC and EggPC mixed with 0.05 and 0.1 mg RP liposomes. The low field line of ^{15}N-PDT spin label ESR spectra was recorded every two min after UVA light was turn on. The ESR signal was recorded with 0.5 mW incident microwave and 100 kHz field modulation of 0.05G at 37C. Lipid peroxidation was initiated by photoexcitation with UVA light (320 mm).*

Based on these combined results, the proposed mechanisms of generation of ROS and induction of lipid peroxidation by photoexcited RP light are shown in Figure 9.

4 CONCLUSION

In a biological system, free radicals are not generally generated in high concentrations and, due to their high reactivity, do not persist. As such, detection of free radicals is difficult. Therefore, it is difficult to study the mechanisms that generate free radicals. In this paper we demonstrate the use of ESR techniques to study the anti-oxidant and pro-oxidant capabilities of RP. These results clearly illustrate the potential application of ESR spectroscopic methodologies for the study of antioxidant and pro-oxidant activities of food extracts, phytochemical constituents, and dietary supplements, and to elucidate the mechanisms underlying the antioxidant and pro-oxidant capabilities via effects on levels of ROS, lipid peroxidation, and damage to or protection of cellular membranes.

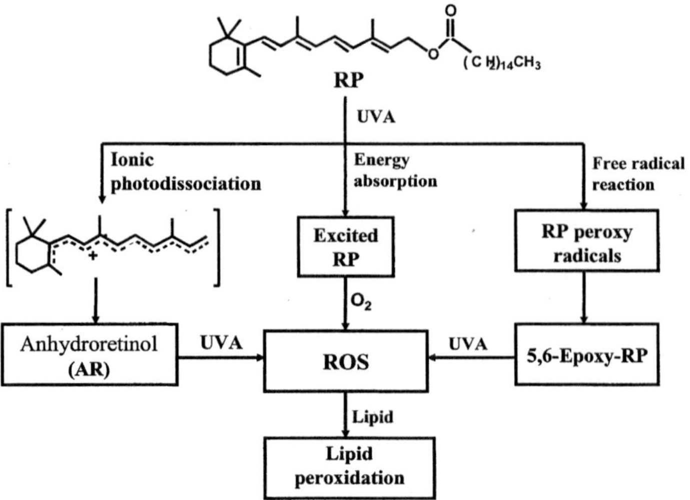

Figure 9 *The proposed mechanisms for generation of ROS and lipid peroxidation following photoexcitation of RP with UVA light.*

References

1   J.J. Yin and P.P. Fu. in: *Magnetic Resonance in Food Science, Challenges in Changing World*, eds. M. Gudjonsdottir, P. Belton and G. Webb, , Royal Society of Chemistry, Cambridge, 2009, p 213.

2   E.R. Stadtman and B.S. Berlett, *Chem. Res. Toxicol.*, 1997, 10, 485.

3   J. Moore, J.J.Yin and L. Yu, *J. Agric. Food Chem.*, 2006, **54**, 617.

4   J.J. Yin, J.K.G. Kramer, M.P. Yurawecz, A.R. Eynard, M.M. Mossoba and L.Yu, *J. Agric. Food Chem.*, 2006, **54**, 7287.

5   J.J. Yin, M.M. Mossoba, J.K.G. Kramer, M.P. Yurawecz, K. Eulitz, K.M. Morehouse and Y. Ku, *Lipids,* 1999, **34**, 1017.

6   J. Xie, Z.H. Shao, T.L. Vanden Hoek, W.T. Chang, J. Li, S. Mehendale, C.Z. Wang, C.W. Hsu, L.B. Becker, J.J. Yin and C.-S. Yuan, *Eur. J. Pharmacol.*, 2006, **532,** 201.

7   J.J. Yin, M.J. Smith, R.M. Eppley, A.L. Troy, S.W. Page and J.A. Sphon, *Arch. Biochem. Biophys.*, 1996, **335**, 13.

8   J.J. Yin, M.J. Smith, R.M. Eppley, S.W. Page and J.A. Sphon, *Biochem. Biophys. Acta-Biomembranes,* 1998, **1371**, 134.

9   K. Zhou, J.J. Yin and L. Yu, *Food Chem.*, 2006, **95,** 446.

10 W.T. Chang, Z.H. Shao, J.J. Yin, S. Mehendale, C.Z. Wang, Y. Qin, J. Li, W.J. Chen, L.B. Becker, T.L. Vanden Hoek and C.-S. Yuan, *Eur. J. Pharmacol.*, 2007, **566,** 58.

11 H.N. Yu, J.J. Yin and S.R. Shen, *J. Agric. Food Chem.*, 2004, **52**, 462.

12 C.Y. Wang, S.Y. Wang, J.J. Yin, J. Parry and L.L. Yu, *J. Agric. Food Chem.*, 2007, **55**, 6527.

13 P. Fu, Q. Xia, M.D. Boudreau, P. Howard, W. Tolleson, and W.Wamer. in *Vitamins and Hormones – Vitamin A, Vol 75*, ed. G. Litwack, Academic Press, Elsevier Inc., San Diego, CA,2007, ch. 9, 223.

14 P.P. Fu, Q. Xia, J.J. Yin, S.-H.Cherng, J. Yan, N. Mei, T. Chen, M.D. Boudreau, P.C. Howard and W.G. Wamer, *Photochem. Photobiol.*, 2007, **83**, 409.

15 J.J. Yin, Q. Xia, S.-H. Cherng, I-W. Tang, P.P. Fu, G. Lin, H. Yu and D.H. Sáenz, *Int. J. Environ. Res. Public Health*, 2008, **5**, 26.

16 Q. Xia, M.W. Chou, J.J. Yin, P.C. Howard, H. Yu and P.P. Fu, *Toxicol. Ind. Health*, 2006, **22**, 147.

17 J.J. Yin, F. Lao, J. Meng, P.P. Fu, Y. Zhao, G. Xing, X. Gao, B. Sun, P.C. Wang, C. Chen and X.-J. Liang, *Mol. Pharma.*, 2008, **74**, 1132.

18 W.H. Tolleson, S.H. Cherng, Q. Xia, J.J. Yin, W.G. Wamer, P.C. Howard, H. Yu and P.P. Fu, *Int. J. Environ. Res. Public Health*, 2005, **2**, 147.

19 S.H. Cherng, Q. Xia, L.R. Blankenship, J.P. Freeman, W.G. Wamer, P.C. Howard, and P.P. Fu. *Chem. Res. Toxicol.*, 2005, **18**, 129.

20. Q. Xia, J.J. Yin, W.G. Wamer, S.H. Cherng, M.D. Boudreau, P.C. Howard, H. Yu and P.P. Fu, *Int. J. Environ. Res. Public Health*, 2006, **3**, 185.

21 Q. Xia, J.J. Yin, S.-H. Cherng, W.G. Wamer, M. Boudreau, P.C. Howard and P.P. Fu, *Toxicol. Lett.*, 2006, **163**, 30.

22 J.J. Yin, Q. Xia and P.P. Fu, *Toxicol. Ind. Health*, 2007, **23**, 625.

Metabolomics

NUTRITIONAL METABOLOMICS AS AN APPROACH TO UNRAVEL METABOLIC HEALTH TRAJECTORY

*Sebastiano Collino[1], François-Pierre J. Martin[1], Sunil Kochhar[1], Serge Rezzi[1]**

[1]BioAnalytical Science, Metabolomics & Biomarkers, Nestlé Research Center, PO Box 44, CH-1000 Lausanne 26, Switzerland
*Corresponding author. E-mail: Serge.Rezzi@rdls.nestle.com *Tel.: +41-21-785-9165, Fax: +41-21-785-9486*

1   INTRODUCTION

Modern nutrition research has promoted the use of metabolomics techniques to gain a deeper understanding of the interactions between nutrition, health, and physiological processes [1]. Metabolomics is a well established analytical approach for the analysis of physiological regulatory processes *via* the quantitative measurement of dynamic metabolic changes of living systems in response to genetic modifications or physiological stimuli, including diet and nutrients (Figure 1.1). Nowadays, metabolomic applications have evolved towards deciphering the cellular and molecular processes in response to different individual dietary modulations, predicting health and disease outcomes. By the global study of low molecular weight compounds in biofluids and tissues, it assures the characterization of individual metabolic phenotypes [2,3]. Metabolomics employs mainly two analytical techniques based on ^{1}H nuclear magnetic resonance (NMR) spectroscopy, and mass spectrometry coupled to gas or liquid chromatography (GC/MS and LC/MS). NMR spectroscopy offers the unique prospect to holistically profile hundreds of metabolites with no *a priori* selection. Interestingly, this technique is not only used for the profiling of biological fluids (liquid state NMR), but it is also today commonly employed for the study of metabolic profiling of intact tissue biopsies, using High Resolution Magic Angle Spinning NMR (HR-MAS) [4]. MS methods are also commonly employed for global and targeted profiling and are inherently more sensitive, but require a more comprehensive sample preparation [5] with separation of the metabolite components using either gas chromatography (GC) or liquid chromatography (LC). Both analytical methods are then comprehensively used to generate high density data, from which meaningful biological information are recovered using advanced statistical tools [6,7].

2   NUTRITIONAL METABOLOMICS: DEPICTING SPECIFIC METABOLIC PHENOTYPES

Metabolomics had displayed in the past numerous applications in diseases diagnostics, and in the investigation of physiological changes caused by toxic insults [8-10]. Yet, nowadays, metabolomics stands as an innovative tool offering a valuable aid to identify disease and prognostic biomarkers which could be implemented in clinical strategies of, for instance, stress-related digestive disorders[11]. Indeed, metabolomics had been shown to be a very powerful instrument to assess the metabolic response to stress [12,13]. There is emerging evidence that chronic and acute stress contributes to the disruption of metabolic homeostasis while influencing individual susceptibility to develop diseases. For example, gastrointestinal diseases such as peptic ulcers and ulcerative colitis are known to be greatly

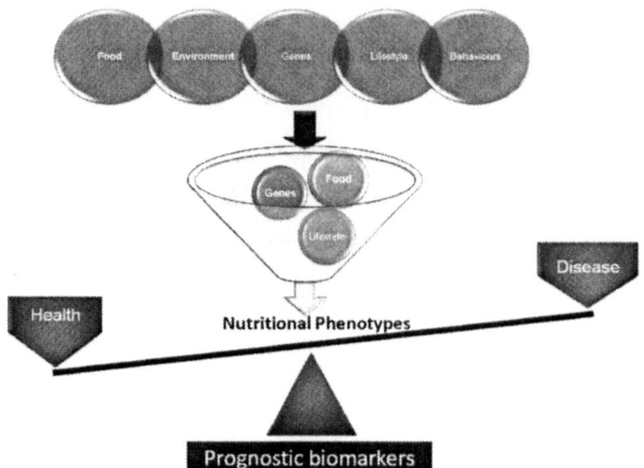

Figure 1 :   *Nutritional metabonomics aims at the understanding the complex metabolic profiles which encapsulate information on genetics, environmental factors, gut microbiota activity, lifestyle, and food habits. Metabotypes are then recovered to discern early biomarkers of homeostasis loss.*

influenced by stress [14].

Although highly debated, epidemiological evidence suggests that life stress is related to the onset, persistence, severity, and prognosis of inflammatory bowel syndrome (IBS) [15,16]. A global view of the metabolic events associated with background stress and its potential influence on the response to novel incoming stress was recently captured in healthy subjects [17]. Stress was displayed to lead to significant perturbations in gut permeability and energy metabolism. Baseline stress was shown to be reflected in the metabolic profile and cold-stress stimulus increased lumen-to-blood passage of small molecules, as a result of impairment of gut permeability. Moreover, levels of ketone bodies, Krebs's cycle intermediates, glucose, and glucogenic amino acids were consistently modulated before and after cold stress applications indicating a stress-induced homeostasis loss.

Other works demonstrated the potential of metabolic profiling for surveying the outcome of nutritional interventions and for modulating or preventing metabolic deregulations [18]. Recently, metabolomics has been applied, in combination with measures of blood plasma inflammatory biomarkers and histopathology of gut tissues, to elucidate the mechanistic basis and biochemical events behind inflammatory bowel disease (IBD) [19]. By following the gradual development of colitis in a Interleukin 10 knockout mice model, with selected spontaneous chronic inflammation, holistic metabolic profiling of blood plasma revealed a gradual disruption of energy homeostasis, profound impairment of lipoprotein, phospholipids and polyunsaturated fatty acids metabolism, and an altered glycosylated protein profile. In addition, IL 10 knockout mice displayed higher levels of lactate, pyruvate, citrate, and higher concentration of free amino acids. All together these metabolic changes indicate increased fatty acid oxidation and glycolysis, with higher level of circulating amino acids reflecting muscle atrophy, increase breakdown of proteins and energy production by interconversion of amino acids.

Yet, in recent years, metabolomics had been also applied to food research applications. Here the term "nutrimetabolomics" entirely describes the mutual link among the fields of metabolomics and nutrition research. Indeed, one of the greatest challenges in modern nutrition is to interrogate and classify the critical metabolic interactions between the complex food matrices, containing a wide range of biologically active compounds, and human system metabolism. Several studies investigated in the past the role of varying nutrient supply on longevity in humans and animal models, and individual responses were strongly dependent on genotype, age, nutrients and regulation of nutrient-sensing pathways [20,21]. Yet, many fundamental questions on aging are still unanswered and/or are object of many intense research efforts. Metabolic profiling is well suited to open new perspectives in understanding the molecular mechanisms of aging at different levels of biological organizations. Within these aims metabolomics has successfully been applied to study the modulation of the ageing processes following nutritional interventions[22,23], including caloric restriction-induced metabolic changes in mouse[24], dogs[25], and non-human primates[26]. Specifically in the canine population, age was associated with increase excretion of creatinine through adulthood, followed by a later decrease within aging in addition to profound changes in gut microbiotal metabolism.

Nutrimetabolomics studies have also revealed that specific dietary metabolic phenotypes imprints, or metabotypes, in both human basal metabolism and gut microbiota activity, are closely related to specific individual dietary preferences [27]. For instance, metabolic profiling of urine revealed that "chocolate liker" is associated to specific energy metabolism and gut microbial activities, thus anticipating possible long term health consequences. Additionally, a combined NMR and MS approach was employed to study the metabolic response due to dark chocolate consumption, with key emphasis on stress related metabolic changes[28]. Here, recruited subjects were previously classified into low and high anxiety using validated psychological questionnaires. Biological fluids were collected at baseline, one week, and two weeks after daily chocolate consumption. Human subjects with higher anxiety trait showed a distinct metabolic profile indicative of a different energy homeostasis (lactate, citrate, succinate, trans-aconitate, urea, proline), hormonal metabolism (adrenaline, DOPA, 3-methoxy-tyrosine) and gut microbial activity (methylamines, p-cresol sulfate, hippurate). Dark chocolate reduced the urinary excretion of the stress hormone cortisol and catecholamines and partially normalized stress-related differences in energy metabolism (glycine, citrate, trans-aconitate, proline, β-alanine) and gut microbial activities (hippurate and p-cresol sulfate). Therefore, subtle changes in dietary habits are likely to modulate the metabolic status that might be associated with long-term health consequences, in particular via the activity of the symbiotic bacterial partners.

3   METABOLOMICS: MEASURING THE CONTRIBUTION OF GUT MICROBIAL METABOLISM TO GUT DISORDERS WHILE DECIPHERING PROBIOTIC'S EFFECTS ON THE MAMMALIAN GUT MICROBIAL METABOLIC CROSS TALK

The recent discovery of the contribution of gut microbiota in the predisposition to gastritis and obesity has raised new interest in gut microbial activities of human and their implications in future nutritional health-care. The gut microbiome-mammalian "Superorganism" [29] represents a level of biological evolutionary development in which there is extensive "transgenomic" modulation of metabolism and physiology that is a characteristic of true symbiosis . The gut microbial community contains multiple cell types providing an extended genome interacting with a number of important mammalian metabolic regulatory functions. As the microbiome interacts strongly with the host to determine the metabolic phenotype [30,31] there is clearly an important role of understanding these interactions as part of personalized healthcare solutions. Many mammalian-microbial associations, both positive and negative, have been also reported [32-34].

The current metabonomic revolution offers an unprecedented opportunity to identify the molecular foundations among the gut and the host relationship so that we can understand how they contribute to normal physiology and how they can be exploited to develop new therapeutic strategies. In the past, conventional techniques have demonstrated that the complex microbial community differs in composition along the length (with an increasing gradient of indigenous microbes from the stomach to the colon) and across the diameter of the gastrointestinal tract, and is comprised of both rapidly transiting and relatively persistent components [35]. The gut symbionts communicate with surrounding tissues to shape a host environment that fosters their implantation and persistence [36]. In addition, the modulation of host-cell pathways by bacteria is reported as a determinant factor in the aetiology of disease and onset of deregulation of metabolic homeostasis [37]. ^{1}H-HR MAS NMR spectroscopy was recently used to characterize the biochemical components of intact epithelial tissues from the different gut compartments along the human gastrointestinal tract (GIT) including the antrum, duodenum, jejunum, ileum, and transverse colon [38]. Regio-specific metabolic variations were observed relating to different physiological properties in each intestinal tissue, including functional variations in energy metabolism, osmoregulation, gut microbial activity, and oxidative protection. Furthermore, as assessed by Martin et al. [39], symbiotic gut microbes can have a significant influence on host health and disease etiology through a region-specific modulation of gastrointestinal functions. The author analyzed the effects of inoculating germfree mice with a model of simplified human baby microbiota (HBM) on the biochemical composition of intact intestinal tissues (duodenum, jejunum, ileum, proximal and distal colon) using magic-angle-spinning ^{1}H NMR spectroscopy. The HBM tissue metabolite profiles with those from conventional and conventionalized mice were compared. Each topographical intestinal region showed a specific metabolic profile that was altered differentially by the various microbiomes, especially for osmolytes. The duodenum and jejunum were characterized with higher concentrations of amino acids, which illustrate the essential role of the small intestine in catabolizing up to 50% of the dietary amino acids for energy production, intestinal *de novo* synthesis and maintenance of mucosa [40]. Notably, conventional mice had higher levels of amino acids in the jejunum, whereas HBM animals showed higher amino acid levels in the duodenum. These microbial-dependent variations indicate that the type of gut bacteria may have important implications for the utilization efficiency of dietary proteins and amino acids and their subsequent availability to extra-intestinal tissues. Recent applications of top-down system biology approached revealed the depth and width of the long range effects of the gut microbiota modulation in complex organisms, resulting in modulation of host lipid, carbohydrate and amino acid metabolism at a panorganismal scale [40,41]. Wikoff et al. have also shown the influence of the microbiome and specific bacteria on mammalian plasma biochemistry [42]. Their application of MS-based metabolic approach revealed significant differences among circulating molecules in blood plasma from germ-free mice and conventional mice. Their work provided additional evidence that the specific metabolic activities of a single gut bacterial species can provide the host with new biochemical compounds in sufficient amount to be detected in the general blood circulation.

Metabolomics is also nowadays a powerful tool effectively used to underlie the mechanistic effects of probiotics on the microbial ecology and on human health. In a recent study, Martin et al. employed such an approach to generate a region-specific metabolic mapping of the gut tract of a gnotobiotic mouse, in which they assessed the effects of *Lactobacillus paracasei* [43]. A total of 24 female gnotobiotic mice were divided into three groups: a control group supplemented with water and two groups supplemented with either live *L. paracasei* or a gamma-irradiated equivalent. Variations in relative concentrations of amino acids, anti-oxidant, and creatine were observed relating to different physiological properties in each intestinal tissue. Metabolic characteristics of lipogenesis and fat storage were observed in the jejunum and colon. Colonization with live *L. paracasei* induced region-dependent changes in the metabolic profiles of all intestinal tissues, except for the colon, consistent with modulation of intestinal digestion, absorption of nutrients, energy metabolism, lipid synthesis and protective functions. These signals acted as reference profiles with which to compare changes in response to gut microbiota manipulation at the tissue level as demonstrated by ingestion of a bacterial probiotic. Contrary to the effects induced by live *L. paracasei*, no changes were seen with supplementation of irradiated-killed bacteria. The results presented here demonstrated the importance of the transgenomic interactions between *L. paracasei* and the host, at the metabolic level to modulate the gut functions.

Additionally, a novel top-down level system biological approach of the host response to probiotic intervention was also revealed, where amino acids metabolism, methylamines, and short chain fatty acids pathways appeared to be altered. Here, the transgenomic metabolic effects of exposure to *Lactobacillus paracei* or *Lactobacillus rhamnous* probiotics were measured by a combined NMR and UPLC-MS metabonomic approach [44]. Biofluids and tissues analysis of germ free mice model, colonized with a simplified model of human microbiota, revealed that probiotic exposure exerted microbiome modification, resulting in altered hepatic lipid metabolism, lowered plasma lipoprotein level, and stimulated glycolysis.

In a follow-up study, Martin et al. employed a similar strategy to monitor biochemical changes across the major compartments of the intestinal tract in response to the inoculation of germfree mice with different microbiomes and subsequent nutritional intervention with probiotics and synbiotics (combination of probiotics and prebiotic) [45]. Colon-specific osmo-regulatory functions for balancing luminal and intracellular milieus were highlighted through higher levels of two osmolytes, *myo*-inositol and *scyllo*-inositol. HBM mice showed a reduction in colon levels of *myo*-inositol and taurine, another strong osmoprotectant, which suggested an alteration of these muscular and osmo-regulatory functions. Proximal and distal colon tissues were differentiated in the different microbiome models by the concentrations of bacterial products, those being the highest in conventional animals. Moreover, supplementing HBM mice with probiotics or synbiotics resulted in different metabolic signatures in the proximal and distal colon, which shows how dietary modulation of bacterial activity may alter host intestinal biochemistry. In another selected mouse model, characterized by simplified gut microbiota mouse system, the depth of the gut microbiome modulations of host biochemistry were also uncovered by modulating the gut functional ecology with pro-, pre-, and synbiotics [45]. Martin et al. assessed the dietary modulations of the gut microbota at a panorganismal scale, by multicompartmental modeling approach with metabolic imprints from 10 biological compartments. Induced microbial changes affected host lipids, carbohydrates, and amino acids metabolism in every major mammalian biological analyzed compartment. Selectively, probiotic supplementation reduced hepatic glycogen, glutamine, and adrenal ascorbate with intrinsic effects on energy homeostasis, antioxidantion, and stereogenesis.

4   CONCLUSION

Many today challenges of nutrition research lie mostly in the ability to decipher the molecular pathways which are affected by nutrients and to assess control over these regulations. Additionally, the ability to evaluate the efficacy of food and specific nutrients on the overall metabolic phenotypes, being influenced by many intrinsic and extrinsic factors, remains a very demanding task. Thorough the rigorous characterization of interactions between the diet and the microbiota, metabolomics is providing new ventures for modulating the microbiota towards the improvement of human health. The prospective of preventing the progression of human diseases, by normalizing their effects, by specific nutritional intervention programs, such as probiotic supplementation, could benefit from the applications of metabolomics for therapeutic surveillance and assessment of treatment efficacy. As the microbiome interacts strongly with the host to determine the metabolic phenotype, metabolic health and nutritional status, there is clearly an important role of understanding these interactions as part of optimized nutrition.

Nutrimetabolomics research is therefore providing a system biological approach that is potentially able to assess metabolic status of individuals considering their specificity in terms of genetic/environmental factors, gut microbiota activity, lifestyle, and food habits.

References

1. S.Rezzi, Z.Ramadan, L.B.Fay, S.Kochhar, *J Proteome.Res,* 2007, **6**, 513.
2. J.K.Nicholson, P.J.Sadler, J.R.Bales, S.M.Juul, A.F.MacLeod, P.H.Sonksen, *Lancet,* 1984, **2**, 751.
3. J.R.Bales, D.P.Higham, I.Howe, J.K.Nicholson, P.J.Sadler, *Clin.Chem,* 1984, **30**, 426.
4. O.Beckonert, m. Coen,H.C. Keun, Y. Wang, T.M. Ebbels, E. Holmes, J.C. Lindon, J.K. Nicholson, 2010, *Nat. Prot.,* **6**, 1019.
5. S.J.Bruce, I.Tavazzi, V.Parisod, S.Rezzi, S.Kochhar, P.A.Guy, *Anal.Chem,* 2009, **81**, 3285.
6. I.Montoliu, F.P.Martin, S.Collino, S.Rezzi, S.Kochhar, *J.Proteome.Res.,* 2009, **8**, 2397.
7. L.Eriksson, H.Antti, J.Gottfries, E.Holmes, E.Johansson, F.Lindgren, I.Long, T.Lundstedt, J.Trygg, S.Wold, *Anal.Bioanal.Chem,* 2004, **380**, 419.
8. M.Coen, S.U.Ruepp, J.C.Lindon, J.K.Nicholson, F.Pognan, E.M.Lenz, I.D.Wilson, *J.Pharm.Biomed.Anal.,* 2004, **35**, 93.
9. B.M.Beckwith-Hall, N.A.Thompson, J.K.Nicholson, J.C.Lindon, E.Holmes, *Analyst,* 2003, **128**, 814.
10. E.Holmes, J.K.Nicholson, A.Nicholls, J.C.Lindon, S.C.Connor, S.Polly, J.Connelly, *Chemom.Intell.Lab.Syst.,* 1998, **44**, 245.
11. J.T.Bjerrum, O.H.Nielsen, Y.L.Wang, J.Olsen, *Nat.Clin.Pract.Gastroenterol.Hepatol.,* 2008, **5**, 332.
12. C.R.Teague, F.S.Dhabhar, R.H.Barton, B.Beckwith-Hall, J.Powell, M.Cobain, B.Singer, B.S.McEwen, J.C.Lindon, J.K.Nicholson, E.Holmes, *J Proteome.Res,* 2007, **6**, 2080.
13. Y.Wang, E.Holmes, H.Tang, J.C.Lindon, N.Sprenger, M.E.Turini, G.Bergonzelli, L.B.Fay, S.Kochhar, J.K.Nicholson, *J Proteome.Res,* 2006, **5**, 1535.
14. J.D.Soderholm, P.C.Yang, P.Ceponis, A.Vohra, R.Riddell, P.M.Sherman, M.H.Perdue, *Gastroenterology,* 2002, **123**, 1099.
15. A.Faresjo, E.Grodzinsky, S.Johansson, M.A.Wallander, T.Timpka, I.Akerlind, *Eur J Epidemiol.,* 2007, **22**, 473.
16. B.I.Nicholl, S.L.Halder, G.J.Macfarlane, D.G.Thompson, S.O'Brien, M.Musleh, J.McBeth, *Pain,* 2008, **137**, 147.

17. S.Rezzi, F.P.Martin, C.Alonso, M.Guilarte, M.Vicario, L.Ramos, C.Martinez, B.Lobo, E.Saperas, J.R.Malagelada, J.Santos, S.Kochhar, *J.Proteome.Res.,* 2009, **8**, 4799.

18. V.P.Makinen, P.Soininen, C.Forsblom, M.Parkkonen, P.Ingman, K.Kaski, P.H.Groop, M.la-Korpela, *Mol.Syst.Biol.,* 2008, **4**, 167.

19. F.P.Martin, S.Rezzi, D.Philippe, L.Tornier, A.Messlik, G.Holzlwimmer, P.Baur, L.Quintanilla-Fend, G.Loh, M.Blaut, S.Blum, S.Kochhar, D.Haller, *J Proteome.Res,* 2009, **8**, 2376.

20. M.D.Piper, W.Mair, L.Partridge, *J.Gerontol.A Biol.Sci.Med.Sci.,* 2005, **60**, 549.

21. D.A.Sinclair, *Mech.Ageing Dev.,* 2005, **126**, 987.

22. F.-P.J.Martin, Y.Wang, N.Sprenger, I.K.S.Yap, S.Rezzi, Z.Ramadan, E.Pere-Trepat, F.Rochat, C.Cherbut, P.van Bladeren, L.B.Fay, S.Kochhar, J.C.Lindon, E.Holmes, J.K.Nicholson, *Mol.Syst.Biol.,* 2008, **4**, 205.

23. S.Rezzi, Z.Ramadan, L.B.Fay, S.Kochhar, *J.Proteome.Res.,* 2007, **6**, 513.

24. C.Selman, N.D.Kerrison, A.Cooray, M.D.Piper, S.J.Lingard, R.H.Barton, E.F.Schuster, E.Blanc, D.Gems, J.K.Nicholson, J.M.Thornton, L.Partridge, D.J.Withers, *Physiol Genomics,* 2006,

25. Y.Wang, D.Lawler, B.Larson, Z.Ramadan, S.Kochhar, E.Holmes, J.K.Nicholson, *J Proteome.Res.,* 2007, **6**, 1846.

26. S.Rezzi, F.P.Martin, D.Shanmuganayagam, R.J.Colman, J.K.Nicholson, R.Weindruch, *Exp.Gerontol.,* 2009, **44**, 356.

27. S.Rezzi, Z.Ramadan, F.P.Martin, L.B.Fay, B.P.van, J.C.Lindon, J.K.Nicholson, S.Kochhar, *J Proteome.Res,* 2007, **6**, 4469.

28. F.P.Martin, S. Rezzi, E.P. Trepat, B. Kamlage, S. Collino, E. Leibold, J. Kastler, D. Rein, L. B. Fay, S. Kochhar, *J Proteome.Res ,* 2009,**8**, 5568.

29. J.Lederberg, *Science,* 2000, **288**, 287.

30. M.E.Dumas, R.H.Barton, A.Toye, O.Cloarec, C.Blancher, A.Rothwell, J.Fearnside, R.Tatoud, V.Blanc, J.C.Lindon, S.C.Mitchell, E.Holmes, M.I.McCarthy, J.Scott, D.Gauguier, J.K.Nicholson, *Proc.Natl.Acad.Sci.U.S.A,* 2006, **103**, 12511.

31. F.P.Martin, M.E.Dumas, Y.Wang, C.Legido-Quigley, I.K.Yap, H.Tang, S.Zirah, G.M.Murphy, O.Cloarec, J.C.Lindon, N.Sprenger, L.B.Fay, S.Kochhar, B.P.van, E.Holmes, J.K.Nicholson, *Mol.Syst.Biol.,* 2007, **3**, 112.

32. J.K.Nicholson, E.Holmes, I.D.Wilson, *Nat.Rev.Microbiol.,* 2005, **3**, 431.

33. R.Ley, P.Turnbaugh, S.Klein, J.Gordon, *Nature,* 2006, **444**, 1022.

34. S.R.Gill, M.Pop, R.T.Deboy, P.B.Eckburg, P.J.Turnbaugh, B.S.Samuel, J.I.Gordon, D.A.Relman, C.M.Fraser-Liggett, K.E.Nelson, *Science,* 2006, **312**, 1355.

35. C.Dunne, *Inflamm.Bowel.Dis.,* 2001, **7**, 136.

36. L.V.Hooper, T.Midtvedt, J.I.Gordon, *Annu.Rev.Nutr.,* 2002, **22**, 283.

37. A.P.Bhavsar, J.A.Guttman, B.B.Finlay, *Nature,* 2007, **449**, 827.

38. Y.Wang, E.Holmes, E.M.Comelli, G.Fotopoulos, G.Dorta, H.Tang, M.J.Rantalainen, J.C.Lindon, I.E.Corthesy-Theulaz, L.B.Fay, S.Kochhar, J.K.Nicholson, *J.Proteome.Res.,* 2007, **6**, 3944.

39. F.P.Martin, Y.Wang, I.K.Yap, N.Sprenger, J.C.Lindon, S.Rezzi, S.Kochhar, E.Holmes, J.K.Nicholson, *J.Proteome.Res.,* 2009, **8**, 3464.

40. G.Wu, *Journal of Nutrition,* 1998, **128**, 1249.

41. F.P.Martin, N.Sprenger, I.K.Yap, Y.Wang, R.Bibiloni, F.Rochat, S.Rezzi, C.Cherbut, S.Kochhar, J.C.Lindon, E.Holmes, J.K.Nicholson, *J Proteome.Res,* 2009, **8**, 2090.

42. W.R.Wikoff, A.T.Anfora, J.Liu, P.G.Schultz, S.A.Lesley, E.C.Peters, G.Siuzdak, *Proc Natl.Acad.Sci.U.S.A,* 2009, **106**, 3698.

43. F.P.Martin, Y.Wang, N.Sprenger, E.Holmes, J.C.Lindon, S.Kochhar, J.K.Nicholson, *J.Proteome.Res.,* 2007, **6**, 1471.
44. F.P.Martin, Y.Wang, N.Sprenger, I.K.Yap, S.Rezzi, Z.Ramadan, E.Pere-Trepat, F.Rochat, C.Cherbut, B.P.van, L.B.Fay, S.Kochhar, J.C.Lindon, E.Holmes, J.K.Nicholson, *Mol.Syst.Biol.,* 2008, **4**, 205.
45. F.P.Martin, N.Sprenger, I.K.Yap, Y.Wang, R.Bibiloni, F.Rochat, S.Rezzi, C.Cherbut, S.Kochhar, J.C.Lindon, E.Holmes, J.K.Nicholson, *J.Proteome.Res.,* 2009, **8**, 2090.

NORMALIZATION IS A NECESSARY STEP IN NMR DATA PROCESSING: FINDING THE RIGHT SCALING FACTORS

F. Capozzi[1,2], A. Ciampa[1], G. Picone[1], G. Placucci[1] and F. Savorani[3]

[1]Department of Food Science, University of Bologna, Piazza Goidanich 60, I-47521 Cesena, Italy, Email: francesco.capozzi@unibo.it
[2]CIRMMP, Via L. Sacconi, 6, I-50019 Sesto Fiorentino (FI), Italy
[3]Department of Food Science, Quality & Technology, University of Copenhagen, Rolighedsvej 30, DK-1958 Frederiksberg C, Denmark

1 INTRODUCTION

Metabolites are the end products of cellular regulatory processes, and their levels can be regarded as the ultimate response of biological systems to genetic or environmental changes. In parallel to the terms 'transcriptome' and 'proteome', the whole set of metabolites synthesized by a biological system constitutes its 'metabolome'.[1]

Conventionally, the analytical procedure can be restricted to the identification and quantification of a selected number of pre-defined metabolites in a biological sample. This process is called metabolite profiling (or, sometimes, metabolic profiling).[2] For example, pre-defined metabolites may belong to a class of compounds (such as polar lipids, isoprenoids, or carbohydrates), or be narrowed down to members of particular pathways. The term metabolite profiling is frequently used in the specific context of drug research in the description of catabolic degradation of an applied chemical.

Conversely, the approach revealing the comprehensive metabolome of the biological system under investigation should be called metabolomics. Metabonomics is a further specialization of the *omics* approach, which is based on the systematic profiling of metabolite levels, but encompasses the systematic and temporal changes of the metabolome in whole organisms occurring in response to diet, lifestyle, environment, etc.[3] Metabonomics must aim at avoiding exclusion of any metabolite by using well conceived sample preparation procedures and analytical techniques.

The resolving power of the chosen analytical method must be high enough to maintain sensitivity, selectivity, matrix independence, and universal applicability. Since data sets for metabonomics are complex by nature, adequate tools are needed to handle, store, normalize, and evaluate the acquired data in order to describe the systemic response of the biological system. Some tools can be adopted by classical structure analysis, such as mass spectrometry (MS) and nuclear magnetic resonance (NMR).[4]

NMR and MS are mostly used to generate global metabolite profiles in metabolomics.[5,6,7] An advantage of NMR is that the samples normally do not require any physical or chemical treatment prior to the analysis, whereas MS usually requires that the metabolites be separated from the sample before detection by liquid chromatography.[8] Although MS is more sensitive than NMR, the latter is a more attractive tool for metabonomics studies because the non-destructive nature of NMR enables observation of the dynamics as well as separation of metabolites in biological samples; in contrast, MS disrupts the structures and the interactions of molecular complexes. Chemometrics, e.g. NMR spectroscopic analysis coupled with multivariate statistical methods, offer a powerful new approach for assessing metabolic function.[9,10]

Pattern recognition and related multivariate statistical approaches can be used to discern significant patterns in complex data sets with the aim of classifying objects by identifying inherent patterns in a set of indirect measurements. Pattern recognition methods, such as principal components analysis (PCA) and classification methods such as partial least-squares discriminant analysis (PLS-DA) and orthogonal projection to latent structures discriminant analysis (OPLS-DA), reduce the dimensionality of complex data sets and thereby facilitate the visualization of inherent patterns in the data accelerating the interpretation.

^{1}H NMR-based metabonomics has been applied in the food sciences,[11] including assessments of green tea,[12] rosemary,[13] honey,[14] grape wine,[15,16] crops,[17,18] etc. Mainly, the spectroscopic data ^{1}H-NMR are analyzed by multivariate data analysis methods to extract information about changes in distinct metabolites of examined samples.

The steps involved in the analysis of metabonomics are as follows:[19,20] (a) post-instrumental processing of acquired spectroscopic data, such as removal of offsets by polynomial baseline correction, calculation of intensity values either on each data point, on each peak, or summed over segmented regions (binning); (b) production of a data table from the analytical measurements such that there are m rows (observations, samples) and n columns (variables, frequencies, integrals); (c) normalization of the data or some related adjustment to the spectral intensities (a row operation); (d) scaling of the data (a column operation); (e) multivariate statistical modelling of the data.

The normalization step can is applied to the data from each sample and comprises methods to make the data from all samples directly comparable with each other.[20,21] One of its common applications is to remove or minimize the effects of variable dilution of the samples. Dilution is defined as a process influencing the concentrations of all metabolites, and thus all peak intensities of the corresponding spectrum, by the same factor (coefficient), which can also be referred to as unspecific changes of metabolites.

In contrast, metabolic responses influence a few metabolites in the sample and, consequently, only a few peaks of the corresponding spectrum. These specific changes are visible as relative changes of concentrations of few metabolites related to the concentrations of all other metabolites, which represent the overall concentration of the sample. Usually these specific relative changes are of interest in metabonomics studies in contrast to the overall concentration of the sample, and are only visible after an adequate normalization of the spectra.

Normalization can also be necessary due to technical reasons. If spectra are recorded using a different number of scans, or if spectra are recorded with different devices, the absolute values of the spectra are different and rendering a joint analysis of spectra without prior normalization is impossible.

The specific object of the present investigation was to find a new algorithm, developed on a dataset of NMR spectra recorded on 20 mixtures of 10 different metabolites, also exerting signal overlap, able to correct the dilution errors that have been purposely and artificially introduced during sample preparation. The developed algorithm was also applied on a "natural" spectral dataset of tomatoes extracts, for which both NMR and HPLC techniques are applied in a non-hyphenated manner.

2 METHOD AND RESULTS

2.1 Samples preparation

2.1.1 The mixtures. Twenty mixtures, each containing glutamic acid, p-OH-benzoic acid, histidine, proline, valine, serine, imidazole, aspartic acid, adenosine 5'-triphosphate, phenylalanine, were prepared in 50 mM phosphate buffer at pH 7, by mixing 10 μl of 50 mM solutions of the first 5 substances, and 5 to 25 μl of 50 mM solutions of the remaining five chemicals. According to this procedure, half metabolites were initially kept at constant concentration in all mixtures, whereas the concentration of the other half metabolites was allowed to vary from 0.5 to 2.5 times the amount of those kept at constant concentration. Subsequently, all mixtures were diluted adding different known volumes of water, from 0.87 to 2.80 ml. From each mixture 800 μl were transferred in a NMR tube for the spectroscopic analysis.

2.1.2 Cherry tomatoes. Eleven different samples of cherry tomatoes have been grinded, freeze-dried and stored under argon atmosphere. Before NMR analysis, 2 g of freeze-dried powder have been extracted with 50 ml of $CHCl_3$, for 48 hours, under magnetic stirring. The suspension has been filtered and concentrated to 1 ml, by solvent evaporation under N_2 flow. A 0.1 ml aliquot of this sample has been diluted to 0.5 ml with $CHCl_3$, for the HPLC analysis, whereas the remaining 0.9 ml aliquot has been completely dried and re-solubilised in 1 ml of $CDCl_3$ 99,8 % + TMS 0.05% for the NMR analysis.

2.2 NMR acquisition parameters

2.2.1 Mixtures samples. 1D ^{1}H- NMR spectra were recorded at 298 K on a Varian Mercury Plus AS400 spectrometer operating at 400.097 MHz. The pulse sequence used contains the presaturation step of the water signal, obtained by centring the spectral window at 4.706 ppm and using a 2 s pulse with an attenuation of 5 dB. The spectral window (SW) was 14 ppm, the number of points of the spectrum were 1 K and 90° proton pulses (9.30 μs at 55 dB attenuation) with 6 s time delay were used. Each spectrum was obtained over 512 scans and the FID, prior to Fourier transformation, was multiplied by an exponential factor (line broadening) equal to 1.5 Hz. Phase and baseline corrections of spectra were manually performed by using the software package MestRe-C 4.9.8.0 (www.mestrec.com, MestRe-C Research SL, Santiago de Compostela, Spain). This software was also used to operate integration of each spectrum by selecting 30 integral regions and to export data in ASCII. Afterwards data were statistically processed by using the free software package *R* version 2.11.1.

2D ^{1}H,^{1}H TOCSY was registered with the standard "mlevphpr" pulse program, with water presaturation during relaxation delay, a spectral width of 6 kHz in both dimensions, a 6 s relaxation delay, and 80 ms mixing time, 1 K data points in F2 and 512 increments in F1. Also in this case, phasing of the two-dimensional spectrum was performed by using the software package MestRe-C.

2.2.2 Cherry tomatoes samples. All 1D ^{1}H-NMR spectra have been recorded using a Varian Mercury Plus AS400 spectrometer, operating at 400.097 MHz ^{1}H Larmor frequency. 1D ^{1}H-NMR experiments have been recorded at 298 K with a 30° pulse (2.1 μs) over a spectral width of 5583.47 Hz (14 ppm), 16K complex data points, 1s relaxation delay, 1024 scans, 20 Hz spinning. Each spectrum has been Fourier transformed using a 0.5 Hz Line Broadening apodization. The phase and baseline correction as well as the chemical shift calibration have been accurately performed over all the spectra by using the software MestRe-C. The same program has been used to export data in ASCII format with

16K data points. Chemical shift referencing has been performed by setting at 7.26 ppm the signal of the residual CHCl$_3$ solvent in all spectra.

A single 2D ^{1}H-^{1}HTOCSY spectrum has been acquired using a Bruker spectrometer operating at 800.1338 MHz equipped with a cryoprobe, under the following conditions: 298°K, TPPI phase sensitive mode, the standard mlevphpr pulse program, 0.213 s of acquisition time, 1K data complex points in F2 and 512 increments in F1, spectral width of 9615.38 Hz, 80 ms mixing time and a relaxation delay of 2.5 s. In order to prevent errors due to sample modifications during storage, a 1D ^{1}H-NMR spectrum of the same sample has been acquired before and after the TOCSY experiment. A comparison of the 1D spectra pre and post TOCSY experiment confirmed that no appreciable molecular modifications happened during the acquisition time. Also the 2D TOCSY spectrum has been processed with MestRe-C.

2.3 HPLC analysis

HPLC analyses have been carried out using an HP 1100 series instrument (Agilent Technologies, Palo Alto, CA), equipped with a binary pump delivery system, a degasser, an autosampler, and an HP diode array UV-vis detector (DAD). Chromatography has been performed with a reverse-phase Luna C18 (Phenomenex, Torrance, CA) column of 5 μm particle size and 25 cm x 3.00 mm i.d. The mobile phase flow rate has been chosen in 0.7 ml·min^{-1}. The acquisition wavelength has been set at 455 nm. The injection volume was 10 μL, which is an aliquot withdrawn from the same samples utilized for the NMR analysis. All of the analyses have been carried out at 303 °K using a thermostatic oven. The gradient elution has been performed using a mobile phase of MetOH/H$_2$O/ACN (80/10/10) pump A; ACN pump B. Gradient profile: from 0% B up to 95% in 60 min, 95% B for 59 min, down to 0% B in 10 min for a total of 129 min of elution time. All the HPLC chromatograms have been processed with LC/MSD ChemStation Rev. A.08.03 (Agilent Technologies) in order to perform a blank subtraction and a baseline correction. The same program has been used to export chromatograms in ASCII format with 19350 data points each one.

2.4 The TOCSY-filter normalization algorithm

The normalization algorithm, that has been developed and tested within this study, executes the following steps (*in italics*) in the R-project programming environment:
1) Chose the cross-correlations to be included in the optimization process (3 false and 1 true)
2) Define the first false cross-correlation as the one between integral regions #16 and #21
```
a1f <- 16; b1f <- 21
```
3) Define the second false cross-correlation as the one between integral regions #21 and #27
```
a2f <- 21; b2f <- 27
```
4) Define the third false cross-correlation as the one between integral regions #20 and #30
```
a3f <- 20; b3f <- 30
```
5) Define the true cross-peak as the one between integral regions #14 and #15
```
aV <- 14; bV <- 15
```
6) `dil` represents the vector of 20 dilution coefficients, initially set equal to 1
```
dil <- rep(1,20)
```
7) Define the optimization function `crosscorr` (`integrals` represents the non-normalized NMR integrals data matrix)

```
Crosscorr <- function(dil)
{p1 <- integrals[,a1f] * dil
q1 <- integrals[,b1f] * dil
regress.f1 <- lm(formula = p1~q1)
r1f <- summary.lm(regress.f1)$r.squared
p2 <- integrals[,a2f] * dil
q2 <- integrals[,b2f] * dil
regress.f2 <- lm(formula = p2~q2)
r2f <- summary.lm(regress.f2)$r.squared
p3 <- integrals[,a3f] * dil
q3 <- integrals[,b3f] * dil
regress.f3 <- lm(formula = p3~q3)
r3f <- summary.lm(regress.f3)$r.squared
pV <- integrals[,aV] * dil
qV <- integrals[,bV] * dil
regress.f6 <- lm(formula = pV~qV)
rV <- summary.lm(regress.f6)$r.squared
filter = r1f + r2f + r3f - rV
return(filter)}
```

`filter` is the objective function obtained by combination of the three false (rNf) and one true (rV) cross-correlation elements characterized by their regression coefficients, R^2, and undergoing minimization through the optimization of the `dil` vector.

```
8)res<-optim(dil, crosscorr, control=list(maxit=20000))
```

`res$par` is the vector containing the calculated dilution coefficients returned from the algorithm.

2.5   The TOCSY-filter normalization algorithm for ^{1}H-NMR metabonomics

The effect of the variable large amount of water in food samples is of paramount importance in metabonomics study, since the consequent dilution errors concerning all metabolites dissolved in their extracts affect the proper data processing. For this reason, the samples are often pre-treated for preservation purposes, but not only, through freeze-drying with quantitative elimination of water, which in most cases, being the main component, can be as high as 90% of weight. An inconsistent and unreliable freeze-drying process, or a bad conservation easily lead to a not reproducible application of the normalization step, and then to a wrong and inconsistent pre-processing of data, which can vary even among different replicates of the same sample.

2.5.1 Evaluation of the normalization algorithm on an artificial set of mixtures. In order to develop an algorithm able to correct these dilutions errors on real matrices, at first data were collected on ^{1}H-NMR spectra acquired on an artificial trial set of twenty mixtures, each one consisting of variable concentrations of the same ten molecules. In each mixture, five substances were always kept at a constant concentration (glutamic acid, p-OH-benzoic acid, histidine, proline, valine), while the remaining five chemicals (serine, imidazole, aspartic acid, adenosine 5'-triphosphate, phenylalanine) were added at different concentrations. The mixtures were then further diluted with appropriate additions of H_2O in order to have dilution coefficients, K, ranging between the maximum value of 5.8, for the more diluted mixture, and the minimum value of 2.0, for the more concentrated one.

The ^{1}H-NMR spectra were calibrated with respect to the chemical shift of the valine methyl signal falling at 0.99 ppm. In Table 1, the 36 signals assigned in the ^{1}H-NMR spectra of the trial mixtures are listed, giving rise to 30 integral regions generating a 20 rows/spectra x 30 columns/integrals data matrix. Six signals (namely 1, 5, 12, 13, 26, 28) are indeed overlapped to other six resonances (1', 5', 12', 13', 26', 28', respectively), and each pair is included in the same integral region.

Table 1 *Resonance assignments with chemical shift and multiplicity of the signals from all metabolites present in the 30 integral regions selected in the ^{1}H-NMR spectra (s – singlet, d – doublet, t – triplet, m – multiplet, dd – double doublet, dt – double triplet). Chemical shifts with the same signal number (still distinguished by the prime sign) refer to signals belonging to different metabolites but falling in the same integral region.*

Signals	^{1}H Shift (ppm)	Molecule	Assigment	Multiplicity
1	8,357	adenosine 5'-triphosphate	2'-CH	s
1'	8,326	imidazole	C2-H ring	s
2	8,107	adenosine 5'-triphosphate	8'-CH	s
3	7,997	histidine	C2-H ring	s
4	7,638	p-OH-benzoic acid	α-CH	d
5	7,231	phenylalanine	C2-H ring	m
5'	7,153	imidazole	C4-H ring	s
6	7,012	histidine	C4-Hring	s
7	6,760	p-OH-benzoic acid	β-CH	d
8	5,963	adenosine 5'-triphosphate	1' ribose	m
9	4,414	adenosine 5'-triphosphate	2' ribose	m
10	4,227	adenosine 5'-triphosphate	3'-4' ribose	m
11	4,070	adenosine 5'-triphosphate	5' ribose	m
12	3,961	proline	α-CH	t
12'	3,945	phenylalanine	α-CH	dd
13	3,835	histidine	α –CH	dd
13'	3,804	serine	β –CH	dd
14	3,726	glutamic acid	α-CH	t
15	3,679	serine	α-CH	dd
16	3,585	aspartic acid	α-CH	dd
17	3,444	valine	α-CH	d
18	3,210	proline	δ-CH	t
19	3,163	proline	δ'-CH	t
20	3,100	phenylalanine	β-CH	dd
21	3,037	histidine	β-CH	dd
22	2,959	histidine	β'-CH	dd
23	2,662	glutammic acid	β-CH	dt
24	2,615	glutammic acid	β'-CH	dt
25	2,521	glutammic acid	γ-CH	t
26	2,177	proline	β-CH	m
26'	2,177	aspartic acid	β-CH$_2$	dd
27	2,099	valine	β-CH	m
28	1,848	proline	γ-CH	m
28'	1,911	aspartic acid	β'-CH$_2$	dd
29	1,160	impurity	?	s
30	0.990	valine	γ, γ' –CH$_3$	d

The aim of the comparison between the STOCSY autocorrelation matrix and the simplified TOCSY array is to verify whether only the signals belonging to the same molecule are correlated with high values of R^2 in the autocorrelation matrix.

The NMR dataset was used to generate the corresponding autocorrelation matrix, whose cross-correlation elements possess coefficients R^2 resulting from the regression analysis among all column pairs representing the 30 integral regions (Figure 1A). The autocorrelation matrix is also known as the statistical total correlation spectroscopy analysis (STOCSY).[22] The highest correlations are represented by black and dark grey dots, with R^2 greater than 0.8 and 0.7, respectively, while the least correlated integral regions are coded in medium and light grey, with R^2 less than 0.5 and 0.4, respectively. The autocorrelation matrix resulting from the original non-normalized spectra is then compared with another array (Figure 1B) that is the simplified binned representation of the 2D-TOCSY spectrum acquired on the most concentrated mixture of metabolites (Figure 2), in a way that the two matrices dimensionally fit. The off-diagonal dots in Figure 1B represent true correlations among protons belonging to the same molecule. The TOCSY spectrum, indeed, permits to identify all the signals of an entire spin system of the molecules dissolved in the mixture. For example, by inspecting the valine signals, we can assign the vicinal proton pairs in α– and β–CH, as well as the γ, γ' methyl groups, with respect to carboxyl group carbon, that fall, respectively, to 3.44, 2.09, and 0.99 ppm.

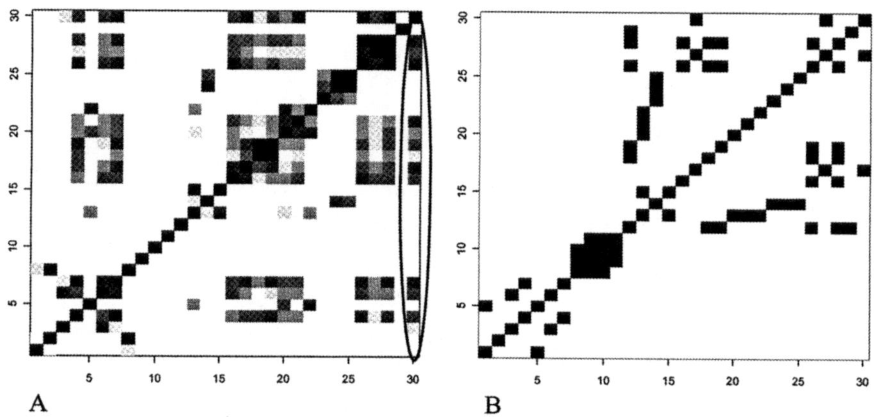

Figure 1 A) *Autocorrelation matrix obtained from the regression analysis among the ¹H-NMR integrals defined for the twenty non-normalized spectra. The integral region #30 originates many cross-correlations (circled with a black line), mostly unexpected.* B) *Data matrix generated from the 2D-TOCSY spectrum. Cross-peak are due to scalar magnetic coupling between signals and are represented by black dots.*

While the 1D proton spectrum provides, through the chemical shift value, information on the chemical nature of all metabolites present in the mixture, the 2D-TOCSY provides also information on which are the signals belonging to the protons of the same molecule. The two-dimensional spectrum can resolve signals of one molecule even though they are overlapped in the 1D spectrum with others belonging to another molecule of the mixture.

Differently from the 2D-TOCSY, which gives only true correlations among signals of the same metabolite, the STOCSY autocorrelation matrix originating from the non-normalized spectra shows also false correlations between signals correctly assigned to different molecules, even with high R^2 values (Figure 1A). For example, signal #30 assigned to valine correlates only with the signals 27 and 17 (having chemical shift 2.09 and 3.44 ppm, respectively), in the 2D-TOCSY (Figure 1B), whilst the same signal correlates with a large number of integral regions, in the non-normalized autocorrelation matrix (Figure 1A): 7 (black – very good correlation), 28, 27, 26, 21, 17, 16, 6 and 4 (dark grey – good correlation), 19 and 20 (medium grey – acceptable correlation) and with signals 18 and 3 (light grey – low correlation).

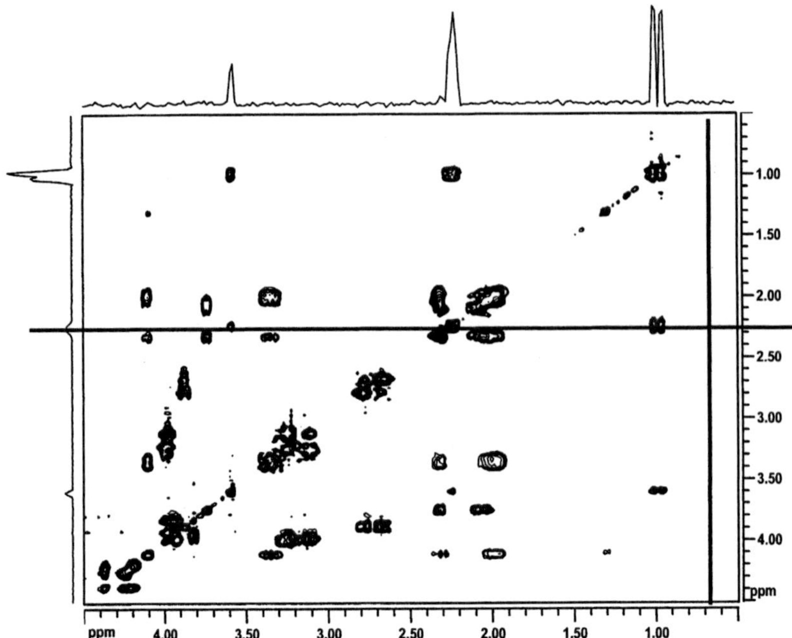

Figure 2 *2D-TOCSY spectrum acquired on the most concentrated mixture. Valine correlations are traced with two lines crossing in correspondence of the off-diagonal peak between the signal at 0,990 ppm assigned to γ,γ'-metyl protons and the signal at 2.099 ppm from β-CH*

In order to verify that the autocorrelation matrix calculated by the mathematical regression would be more similar to the spectroscopic 2D-TOCSY matrix, if the spectra were correctly normalized, each of all 20 spectra has been multiplied by the corresponding known dilution factor K before proceeding with the calculation of the autocorrelation matrix. In this way, the autocorrelation matrix shown in Figure 3A is obtained. By comparing this matrix with the one obtained by using non-normalized spectra (Figure 1A), it is possible to appreciate how the normalizing operation reduces the number of false correlations among integrals.

Except for integrals 27 and 17 (true correlations) all the others are due to the artefacts derived by a wrong dilution operation occurring during samples preparation.

In this case, only one correlation with the integral region #17 was obtained for integral region #30. Thus, the new matrix is almost characterized by the presence of only true correlations as in the 2D-TOCSY. Such a good result was achieved because the dilution factors were previously known since they were intentionally obtained during samples preparation. In true systems, the dilution factors for a correct normalization are unavailable and a proper way to calculate them is desirable.

In order to calculate the actual dilution K-factors, an algorithm has been developed in the R-project environment, namely the "TOCSY-filter algorithm". This algorithm has the purpose to minimize the number of false correlations that may be present in a non-normalized matrix but not confirmed by the TOCSY spectrum. To apply the algorithm, three false and one true cross-correlations, all with R^2 values higher than 0.8, were selected. The algorithm minimize R^2 corresponding to the false cross-correlations and, simultaneously, maximize R^2 corresponding to the true cross-correlation, by multiplying each spectrum by an opportune dilution K-factor. This operation is repeated several times, with different pools of false and true cross-correlations, each trial providing a K-vector consisting of 20 dilution coefficients.

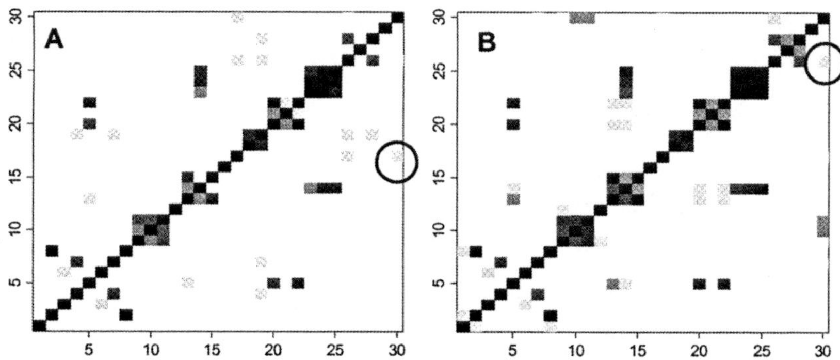

Figure 3 A) *Autocorrelation matrix obtained by multiplying the spectra of all 20 mixtures by the corresponding dilution factor K (normalization). The result is a decrease number of false correlations among integrals, as it can be seen from the number 30.* B) *Autocorrelation matrix obtained by multiplying the spectra of all 20 mixtures by the corresponding dilution factor found out by the algorithm developed in the present study.*

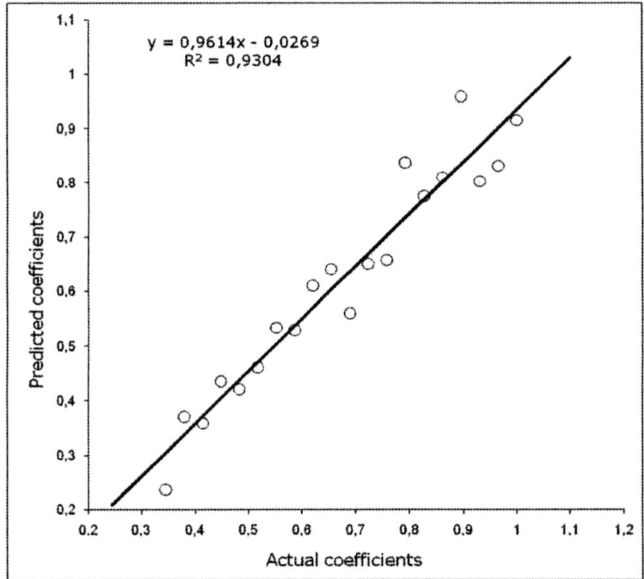

Figure 4 *Correlation between the calculated K-dilution coefficients and the actual dilution factors, that are known because artificially introduced during the preparation of the twenty mixtures.*

At the end, a mean K-vector is calculated by averaging each term over all the dilution vectors obtained by the different trials. Before calculating the mean calculated dilution coefficients, all vectors are preliminarily normalized by dividing all terms in the K-vector by the highest internal K value. In Figure 4 all the calculated K-coefficients (on the average of 6 trials) were compared with the actual ones. The result is a good linear correlation among the actual and the calculated K-factors.

The efficacy of the TOCSY-filter, is readily appreciable by looking at the new autocorrelation matrix (Figure 3B) which is obtained by multiplying the 20 raw spectra by the corresponding K-factors calculated by the algorithm.

Comparison among Figures 1A, 3A and 3B, demonstrates that the algorithm improves the reliability of the autocorrelation matrix. In fact, by considering once again the signal #30, it is possible to find the correlation dots with signals #17 and #27, that are considered as true correlations on the basis of the 2D-TOCSY spectrum.

The algorithm, in some cases, has also shown the capability to work with overlapped signals. For example, in the integral region #13 two signals are present (13 and 13'): at 3.835 ppm for histidine and at 3.804 ppm for serine (see Table 1). The serine correlates with signal #15 at 3.679 ppm, whilst histidine is connected with signal #21 at 3.037 ppm and signal #22 at 2.959 ppm. The algorithm did not fail to find correlations between integral region #13 and signal #15 for serine and between integral region #13 and signal #22 for histidine.

2.5.2 Application of the normalization algorithm on a real food system. A spectral dataset processed with a wrong normalization method produces unreliable statistical results, since the meaningful variance of the metabolite levels among all samples may be hindered by large differences of solvent quantity. Such a variability of the solutes/solvent proportions among the samples determines also the poor co-variance between the observables of a molecule when measured by different analytical techniques on the same samples set. The co-variance analysis is the basis of the statistical hetero-spectroscopy (SHY), which is an extension of STOCSY.[23] In principle, a co-variance matrix between NMR and HPLC data will show correlations among the chemical shift of NMR signals and the retention times of the corresponding metabolite in the chromatogram (Figure 5).

For the biological extracts, the co-variance matrix has a large number of cross-correlations between NMR signals and chromatographic peaks (Figure 5). The number is however much higher than the expected and, as for STOCSY, this is diagnostic for the presence of false correlations associated with a wrong normalization procedure. In the co-variance matrix, false correlations are due to different dilutions of the samples, like those generated by different amounts of residual water in the lyophilized samples' powders. Such variations may also be induced by some inevitable sampling errors, especially in case of non homogeneous food matrices. This is the case of freeze-dried cherry tomatoes powders, where it is possible to collect, by chance, different anatomic parts of the fruit, such as seeds, peel or pulp with different contents of extractable matter.

Consequently, in order to minimize the dilution effect on NMR spectra, the TOCSY-filter algorithm has been applied on the ¹H-NMR autocorrelation matrix obtained from spectra acquired on 11 cherry tomatoes extracts. The resulting STOCSY matrix (Figure 6A) has been compared with a 2D ¹H-TOCSY spectrum acquired on a single chloroform extract of cherry tomato (Figure 6B), in order to identify true and false cross-peaks in the autocorrelation matrix.

The comparison of the two matrices points out the presence of false cross peaks, for example those circled with a red line in Figure 7. The application of the TOCSY-filter algorithm minimizes the presence of false cross peaks in the autocorrelation matrix (data not shown), by applying 11 dilution K-factors, one for each mixture.

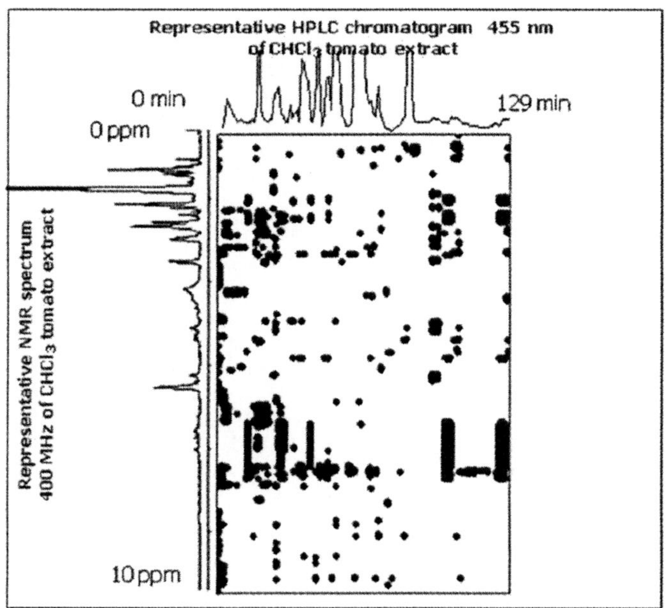

Figure 5 *HPLC-NMR co-variance matrix calculated for non-normalized spectra and chromatograms of natural samples, consisting of chloroform extracts from cherry tomatoes*

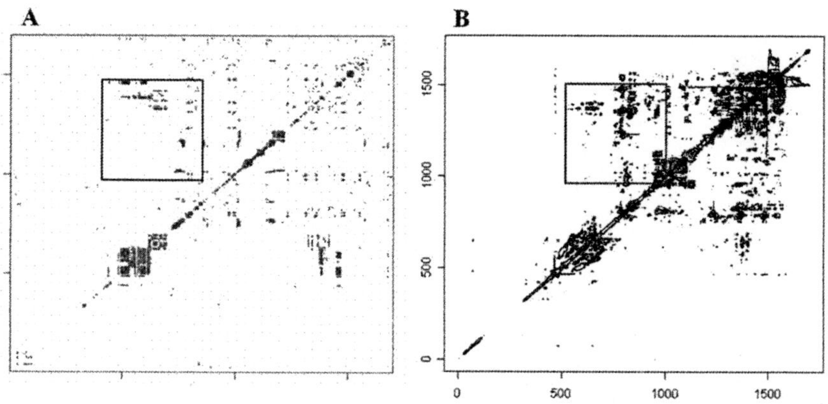

Figure 6 A) *Non-normalized autocorrelation array ^{1}H-NMR (1700 x 1700) showing cross-correlations with $R^{2} > 0.8$.* **B)** *2D-^{1}H-TOCSY, acquired at 800 MHz, on a cherry tomato sample chosen .*

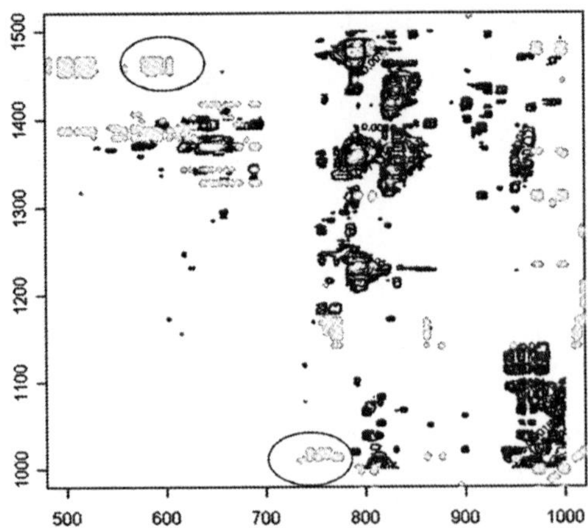

Figure 7 *Superimposition between non-normalized autocorrelation array with only true correlation, in grey, and ^{1}H-TOCSY acquired on a cherry tomato sample, in black. Some false cross peaks are circled.*

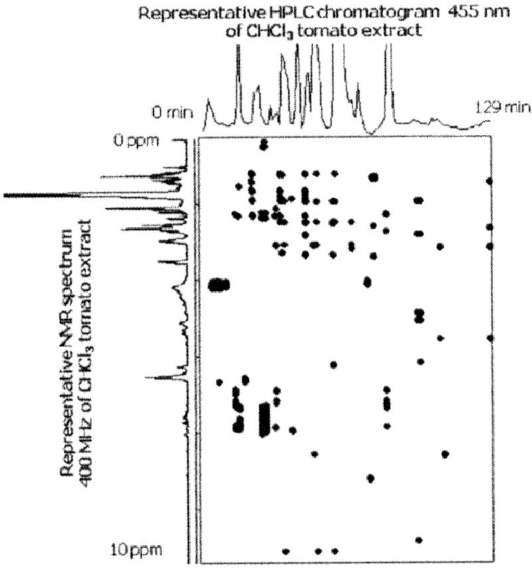

Figure 8 SHY NMR-HPLC *co-variance matrix between the ^{1}H-NMR signals and the chromatogram recorded at 455 nm on the same sample set, showing a minimized number of false cross peaks. The optimization has been performed by applying the TOCSY-filter algorithm.*

Once the dilution K-factors are calculated, they have been used as normalization coefficients for the NMR spectra, as well as for all chromatograms. The resulting SHY co-variance matrix shows the capability of the algorithm to clean out the false cross-correlation, thus improving the reliability of a non-hyphenated HPLC-NMR coupling, that can be used to assign a structure to a molecule whose peak is eluted at the corresponding retention time (Figure 8). It is worth noting that the TOCSY-filter algorithm minimizes the R^2 of artefacts whilst does not affect or maximizes the regression coefficients of true cross-correlations. Thus, the filtering out of a large number of "signals" in the 2D covariance matrix, is not corresponding to the simple cut off procedure obtained by increasing the threshold for the elements to be shown in the image.

Although several resonances are readily observed between 6 and 7 ppm, consistent with the presence of lycopenes in the tomato extracts, not all the main chromatographic peaks, neither all the main NMR signals, shows co-variance between the two techniques. This is easily explained by the assumption that not all the peaks and the signals are resolved, and the overlap among signals and, more frequent, among peaks, is critical for the obtainment of clear covariance. The enhancement of the SHY matrix relies on the improvement of the HPLC separation and of the NMR resolution. However, whilst the latter parameter is constant for a given magnetic field, and can be increased at a larger cost, the first technique can be easily improved by changing the elution gradients and the column performances.

3   CONCLUSION

The TOCSY-filter algorithm, at first developed on a training dataset consisting of NMR spectra acquired on 20 mixtures of 10 different metabolites, has been shown to be able to correct the dilution errors that have been artificially and purposely introduced during sample preparation. Dilution errors is proven to lead to an incorrect normalization of the proton spectra, which is a necessary step prior the multivariate analysis.

The algorithm requires the acquisition of a two-dimensional ^{1}H- TOCSY spectrum on a single sample, chosen among the whole set that will undergo the chemometric analysis. The TOCSY-filter is a simple algorithm, suitable for a future inclusion in the standard procedures so far adopted in metabonomics, as a preliminary step before statistical analysis. The resulting elimination of the effects of incorrect dilution procedures as source of errors, determines an improvement in the discovery of the biological source of variance, which is the main goal in metabonomics, including the applications in foodomics science.

In this field, further information can be gained from covariance analysis of data obtained by two different techniques, either spectroscopic and chromatographic, on the same sample set, in the so called SHY applications. However, when dilution errors are larger than, or comparable to, the biological variance of metabolites concentrations, the co-variance information is poor and the SHY approach is vain. The elimination, or reduction, of the dilution errors, obtained by application of the TOCSY-filter algorithm, may improve the SHY applications, rendering the approach a valid alternative to the hyphenated HPLC-NMR analysis.

References

1   Fiehn, O. *Plant Mol. Biol.* 2002, **48**, 155-171.
2   R.N. Trethewey, A.J. Krotzky and L. Willmitzer, 1999, *Curr. Opin. Plant Biol.,* **2**, 83.
3   J.C. Lindon and J.K. Nicholson, *Ann. Rev. Anal. Chem.,* 2008, **1**, 45.
4   W.M.T. Fan, 1996, *Prog. Nucl. Magn. Reson. Spectrosc.,* **28**, 161.
5   E.M. Lenz and I.D. Wilson, 2007, *J. Proteome Res.,* **6**, 443.
6   J.L. Wolfender, S. Rodriguez and K. Hostettmann, 1998, *J. Chromatogr. A,* **794**, 299.
7   H. Dai, C. Xiao, H. Liu and H. Tang, 2010, *J. Proteome Res.,* **9**, 1460.
8   J. K Nicholson and J. C. Lindon, 2008, *Nature,* **455**, 1054.
9   E. Holmes and H. Antti, 2002, *Analyst,* **127**, 1549

10   A.M. Weljie, J. Newton, P. Mercier, E. Carlson and C.M. Slupsky, 2006, *Anal. Chem.*, **78** , 4430.

11   D.S. Wishart, 2008, *Trends Food Sci. Technol.*, **19**, 482.

12   L. Tarachiwin, K. Ute, A. Kobayashi and E. Fukusakii, 2007, *J. Agric. Food Chem.*, **55**, 9339.

13   C.N. Xiao, H. Dai, H.B. Liu,Y.L. Wang and H.R. Tang, 2008, *J. Agric. Food Chem.*, **56**, 10142.

14   J.A. Donarski, S.A. Jones and A.J. Charlton, 2008, *J. Agric. Food Chem.*, **56**, 5451.

15   H.S. Son, K.M. Kim, F. Van den Berg, G.S. Hwang, W.M. Park, C.H. Lee and Y.S. Hong, 2008, *J. Agric. Food Chem.*, **56**, 8007.

16   G.E. Pereira, J.P. Gaudillere, C. Van Leeuwen, G. Hilbert, M. Maucourt, C. Deborde, A. Moing and D. Rolin, 2006, *Anal. Chim. Acta*, **563**, 346.

17   A.P. Sobolev, A.Segre and R. Lamanna, 2003, *Magn. Reson. Chem.*, 41, 237.

18   H.P.J.M. Noteborn, A. Lommen, R.C. Van der Jagt and J.M. Weseman, 2000, *J. Biotech.*, **77**, 103.

19   R. Goodacre, D. Broadhurst, A.K. Smilde, B.S. Kristal, J.D. Baker, R. Beger, C. Bessant, S. Connor, G. Capuani, A. Craig, T. Ebbels, D. B. Kell, C. Manetti, J. Newton, G. Paternostro, R. Somorjai, M. Sjöström, J. Trygg, F. Wulfert, 2007, *Metabolomics*, 3, 231.

20   A. Craig, O. Cloarec, E. Holmes, J.K. Nicholson and J.C. Lindon, 2006, *Anal. Chem.*, **78**, 2262.

21   F. Dieterle, A. Ross, G. Schlotterbeck and H. Senn, 2006, *Anal. Chem.*, **78**, 4281.

22   O. Cloarec, M.E. Dumas, A. Craig, R.H. Barton, J. Trygg, J. Hudson, C. Blancher, D. Gauguier, J.C. Lindon, E. Holmes and J. Nicholson, 2005, *Anal. Chem.*, **77**, 1282.

23   D.J. Crockford, E. Holmes, J.C. Lindon, R.S. Plumb, S. Zirah, S.J. Bruce, P. Rainville, C.L. Stumpf and J.K. Nicholson, 2006, *Anal. Chem.*, **78**, 363.

TOWARDS IDENTIFICATION OF POLYPHENOL METABOLITES IN BIOFLUIDS
BY SPE-LC-MS-SPE-NMR

M. Klinkenberg[1,3], N. de Roo[1,3], P. Alexandre[1], I. Mahlous[1], D.M. Jacobs[1,3], H.-G.
Janssen[1,3], J. van Duynhoven[1,2,3]

[1]Unilever R&D, Vlaardingen, The Netherlands
[2]Wageningen University, Wageningen, The Netherlands
[3]Netherlands Metabolomics Centre, Leiden, The Netherlands

1 INTRODUCTION

Polyphenols are an important class of functional ingredients that are currently widely being
investigated for potential health benefits. The extensive bioconversion of polyphenols by
colonic microbiota and the host produces a large range of metabolites that can potentially
be taken up into systemic circulation[1,2]. These metabolites are part of the so-called "food
metabolome"[3],[4] which in turn interacts with the human host. Effects on the food
metabolome have typically been assessed by targeted approaches[5]. These, however, hold
the risk of overlooking the compositional complexity of dietary polyphenols and their
metabolic fate in the human superorganism[3]. NMR-profiling has successfully been applied
to capture the food metabolome, with particular focus on gut microbiome-mediated
metabolites[6]. More sensitive and selective GC-[7] and LC-MS[4] profiling approaches have
been developed to capture less abundant food metabolites. These MS-based profiling
approaches have been applied to plasma, urine, feces and in-vitro models, hence capturing
a major part of the microbial bioconversion products of polyphenols. Many of the observed
features in NMR- and MS-based profiles are not known and their identification is a
considerable challenge. In many cases one can take recourse to spectral databases (NMR or
MS), but often one needs to embark on a de-novo molecular identification. This requires
isolation and enrichment of the metabolite(s) of interest and subsequent structural
elucidation by a combination of both NMR and MS[8]. An efficient metabolite identification
platform is provided by combining high-performance LC with NMR and MS. The recent
integration of on-line solid-phase extraction (SPE) with NMR enabled recording of spectra
of metabolites occurring at low concentrations in complex mixtures[9]. A further gain in
sensitivity is provided by the use of cryogenic NMR probeheads. LC-MS-SPE-cryoNMR
platforms have already been deployed for identification of metabolites from single drug
compounds dosed at high levels[10,11]. Nutritional polyphenol formulations are much more
complex and dosages are mostly low. As a consequence, concentrations of polyphenol
metabolites in body fluids such as urine and blood are low. This will in particular pose a
challenge for detection and identification by NMR. Furthermore, low polyphenol
metabolite levels occur against a high background of other (endogenous) metabolites, thus
putting a stronger demand for separating power. Hence we needed to design a strategy for

identification of low-abundance exogenous polyphenol metabolites against the highly complex metabolic background of biofluids. This strategy should fulfil a range of requirements:

1. Ability to obtain pure NMR and MS spectra
2. Sufficient signal-to-noise for metabolites occuring at physiological concentrations
3. Ability to trace back these spectra to features of interest in complex NMR/MS profiles

In order to meet these requirements we optimized concentration, separation and NMR/MS detection in a hyphenated SPE[1]-LC-MS-SPE[2]-NMR approach. This approach was first implemented and tested for identification of metabolites that are being produced upon microbial fermentation of green tea in a colonic in-vitro model. Subsequently, the strategy was further refined for polyphenol metabolites in urine, where levels are much lower and metabolic background is much more pronounced.

2   METHODS

2.1. High Performance Liquid Chromatography

Analyses were carried out using an Agilent 1200 series (Agilent, Waldbronn, Germany) equipped with a G1311A model quaternary pump, a G1322A model vacuum degasser, a G1329A model standard auto sampler and Hystar software (version 3.2; Bruker, Bremen, Germany). The HPLC separations were performed at room temperature by a PRONTOSIL C-18 column (125*4.0mm i.d.; 5.0µm particle size). For the gradient elution, A (acetonitrile + 0.2% acetic acid) and C (water + 0.2% acetic acid) were used as mobile phases; and the total flow rate was maintained at 1 ml/min throughout the run. A Bruker DAD was used for UV detection of polyphenols and was set to an acquisition range of 200-600nm. Two different wavelengths were set for the acquisition: 220nm and 275-285nm.

2.2. Mass Spectrometry

The HPLC system was connected to a micrOTOF (Time-of-Flight) mass spectrometer (Bruker, Bremen, Germany) with Atmospheric Pressure Ionization (API), equipped with an Electrospray Ionization source (ESI) and connected to the LC system by a BNMI-HP unit. MicrOTOF control software (version 2.3 patch 1; Bruker, Bremen, Germany) was used for data acquisition and DataAnalysis software (version 4.0; Bruker, Bremen, Germany) for processing. Ionization was used in electrospray negative mode (API-ES), with a capillary voltage of 3400V, a drying gas flow of 8L/min, nebulizer pressure of 2 bars, drying gas temperature of 200 °C and a charge ratio (m/z) from 50 to 950

2.3. SPE[1] and SPE[2] Cleanup/Concentration

SPE[1] cleanup and concentration by 3cc Oasis HLB cartridges (Waters) was performed in an automated manner by a Gerstel Multi Purpose Sampler (MPS), combined with the Maestro software (version 1.3.7.69; Gerstel, Mülheim an der Ruhr, Germany). The SPE[1] and SPE[2] procedures were based on a protocol for fractionation of polyphenols[12] and further optimized for urine[13]. SPE[2] concentration was performed with a Prospekt 2 (Spark, Emmen, Holland) including an ACE (Automatic Cartridge Exchange) system, a HPD Dual device (High Pressure Dispenser) and HLB Oasis cartridges (Water, 2*10 mm). The cartridges were conditioned with 500 µl of methanol followed by 500 µl of deionised water prior to trapping. After trapping, the cartridges were dried for 30 minutes under nitrogen, in order to remove HPLC solvent. The transfer of compounds from cartridges to NMR tubes was performed with a Gilson Sample Preparation table equipped with PrepGilsonST software (version 1.2; Bruker, Bremen, Germany). Concentrated fractions were eluted from the SPE[2] cartridges with 290μl of deuterated methanol and automatically transferred to NMR tubes through the Prep Gilson robot with a flow of 250μl/min. Mixing was also done by the robot in order to avoid any concentration gradient in the NMR tubes. The residual methanol quintuplet at 3.3 ppm was used as an internal standard for referencing and quantification[14].

2.4. NMR

NMR spectroscopy experiments were performed on a Bruker Avance III 600 NMR spectrometer, operating at 600.23MHz for ^{1}H and equipped with a cryo-cooled probe head and using 3mm NMR tubes. The temperature of the sample was set to 300K. For samples related to SPE[1] optimization, 1D ^{1}H NMR spectra were recorded with the noesypresat or a double irradiation sequence. For noesypresat experiments, presaturation was done during relaxation delay and mixing time. 128 scans were collected in 32K data points with a relaxation time delay of 3.0 s, a mixing time of 150 μs and an acquisition time of 1.36 s. For double irradiation experiments, presaturation was peformed during the relaxation delay. 256 scans were collected in 32K data points with a relaxation time delay of 4.0 seconds, and an acquisition time of 1.36 seconds. A longer relaxation delay of 12s was used for quantitative analysis, in order to have almost full relaxation for most metabolites. The data were processed in TOPSPIN software (version 1.3.6; Bruker BioSpin GmbH, Rheinstetten, Germany). An exponential window function was applied to the free induction decay (FID) with a line-broadening factor of 1.0Hz prior to the Fourier transformation, and phase correction and baseline correction were applied manually.

3   RESULTS AND DISCUSSION

3.1 Design of SPE[1]-LC-MS-SPE[2]-NMR identification strategy

In order to identify polyphenol metabolites in biofluids in a systematic manner, critical steps in the SPE[1]-LC-SPE[2]-MS-NMR approach (Figure 1) needed to be optimized. For this purpose a cocktail of phenolic acids and polyphenols was composed.

Concentration & cleanup: A first SPE[1] step was deemed necessary to obtain a relatively clean and concentrated sample for separation using an analytical LC column. 3 cc HLB cartridges allowed for selective trapping of relatively large amounts of polyphenols, which could be eluted by methanol with high recoveries (>80%).

Separation and UV/MS detection: The LC separation was optimized for the polyphenol cocktail. The chosen analytical LC column allowed injections up to 60 μl before separation was compromised. Both MS TOF and UV were used to continuously monitor the LC elution. The MS TOF was used in negative mode and at the wavelengths of the UV DAD detector polyphenols could readily be observed.

Concentration & NMR detection: Of several commercially available options HLB cartridges turned out to provide the best overall recovery of the polyphenol cocktail. Whereas the commonly used SH cartridges recovered only 3 out of 12 polyphenols, the HLB cartridges had and average recovery of app. 80%. Hence HLB cartridges were used for trapping LC fractions throughout this study. The trapped compounds were eluted with deuterated methanol and collected in 3 mm NMR tubes for off-line NMR measurements in a 5 mm cryoprobe. A modest number of 128 scans was used, resulting in a Limit of Detection at the µM level.

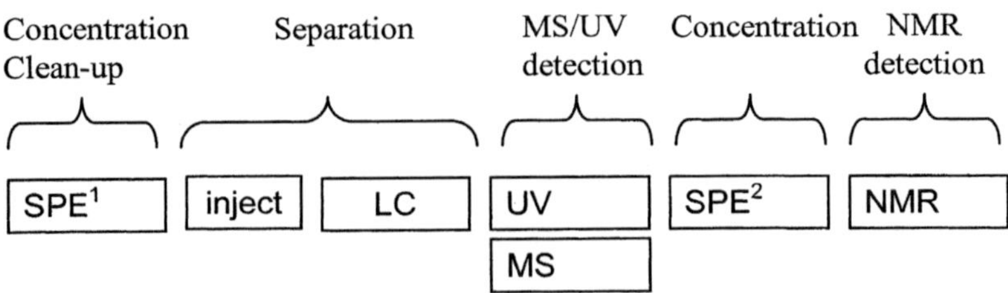

Figure 1 *Schematic depiction critical steps in identification by means of SPE[1]-LC-MS-SPE[2]-NMR*

3.2 Implementation for polyphenol metabolites in in-vitro colon microbiota model

Expected concentration levels of polyphenol metabolites in in-vitro colon models are in the 10-100 µM range and background of other metabolites is not excessive[1]. Hence the SPE[1] clean-up/concentration step was kept rather simple and involved a ten-fold concentration. The injection volume used for the subsequent LC separation was a modest 20 µl and the eluent gradient optimized for model phenolic acids was used. UV and MS settings were taken as optimized for the model mixture, 3 SPE[2] trappings were used for concentrating LC peaks and a modest number of 128 NMR scans was accumulated. Considering a Limit of Detection of 10 µM for NMR and concentration steps and recoveries as determined for model mixtures one can estimate that metabolite levels of 5 µM in the in vitro gut model mixtures can be detected.

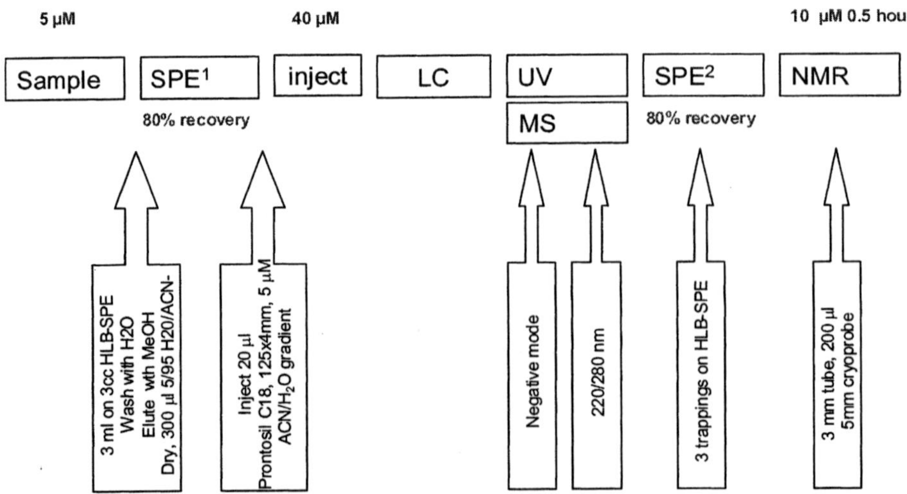

Figure 2 *Optimized procedure for identification of polyphenol metabolites formed by bioconversions in in vitro colon models.*

Typical NMR spectra obtained by 3 trappings of LC peaks are presented in Figure 3. One can observe a signal-to-noise ratio that is sufficient for structural elucidation by NMR.

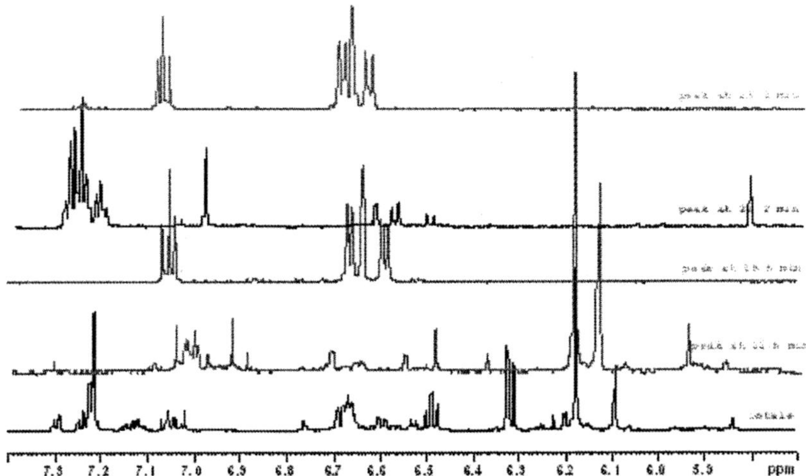

Figure 3 *Examples of NMR spectra obtained by SPE^1-LC-SPE^2 (Figure 2) applied to in vitro gut models after 71 hours of fermentation.*

Structures of gut microbial degradation products of green tea catechins formed at increasing fermentation times are indicated in Figure 4. After 2 hours of fermentation one observes results of scission of the B ring of the EGC, EG, EC and E catechins. After 7 hours one observes formation of 2 valerolactones as result of further microbial degradation. Ultimately (71 h), microbial degradation results in a single-ring phenolic acid.

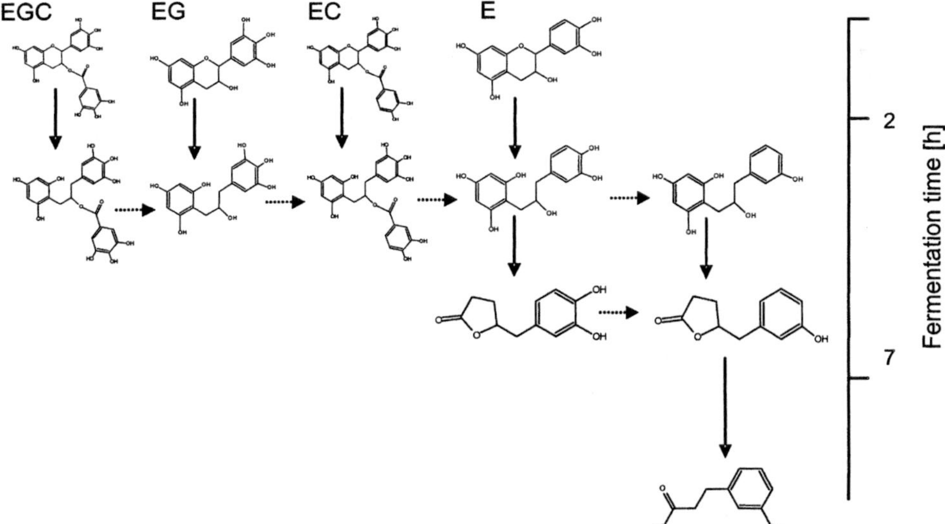

Figure 4. *Possible gut microbial degradation pathways of catechins (EGC, EG, EC, E) present in green tea, which were digested in an in-vitro colon model.*

3.3 Implemenation for polyphenol metabolites in urine

In urine one expects levels of gut-mediated polyphenol metabolites that are in the μM concentration range. These metabolites occur against a complex background of endogenous metabolites. This necessitates a more thorough SPE[1] clean-up/concentration step as was previously applied to model mixtures and in vitro model samples. The injection volume for LC was tripled to 60 μl in order to detect lower metabolite levels and elution gradients were further optimized for urine. The number of SPE[2] trappings was increased to 5, and as a consequence one can detect μM levels in urine (Figure 5).

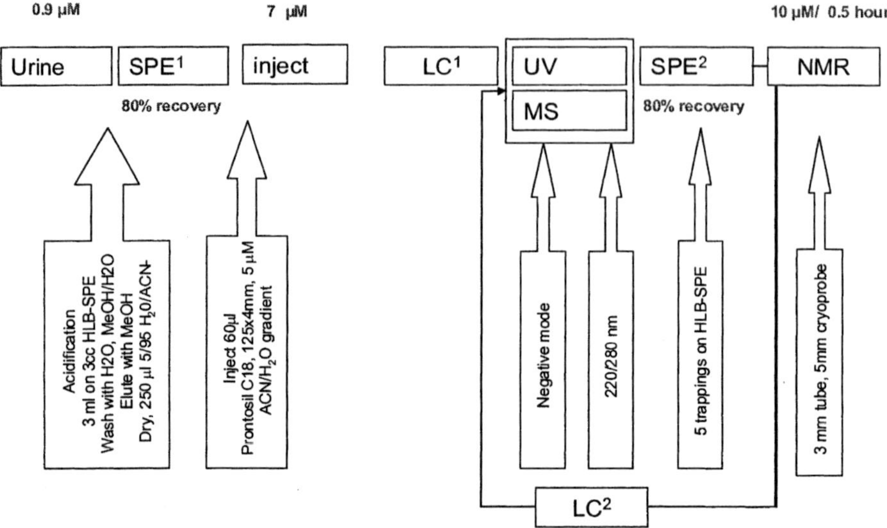

Figure 5 *Optimized procedure for identification of polyphenol metabolites in urine.*

The chromatogram featured many overlapping peaks and was divided in small sections based on the MS trace. These sections were trapped fivefold on the SPE[2] cartridges, NMR spectra of the corresponding elutes showed good signal to noise (Figure 6).

In order to verify overall recovery of injected metabolites by LC-SPE[2] Figure 7 compares the NMR spectra of the SPE1 elute with the co-added spectrum of the respective SPE[2] elutes. Due to different recoveries of the different metabolites by SPE[2] one can observe signal intensities differences, but these are still within the same order of magnitude. Furthermore, spectral positions are well aligned, which enables to link spectral shifts in the initial overall NMR-profile with those in the respective LC-SPE[2] fractions.

Several of the LC-SPE[2] NMR spectra featured major spinsystems that could be identified with the aid of the corresponding MS spectra. NMR and MS spectra of several identified structures are indicated in Figure 7. Of particular interest are the sulphonated and glucuronidation forms of a valerolactone. This indicates that an early gut microbial degradation product of polyphenols undergoes extensive conjugation within the human host. Note that site of conjugation has not been identified yet, this requires information to be provided by ^{13}C NMR. The current signal-to-noise of the 1D ^{1}H NMR spectra acquired in 256 scans would allow for recording good quality HSQC spectra within a reasonable measurement time.

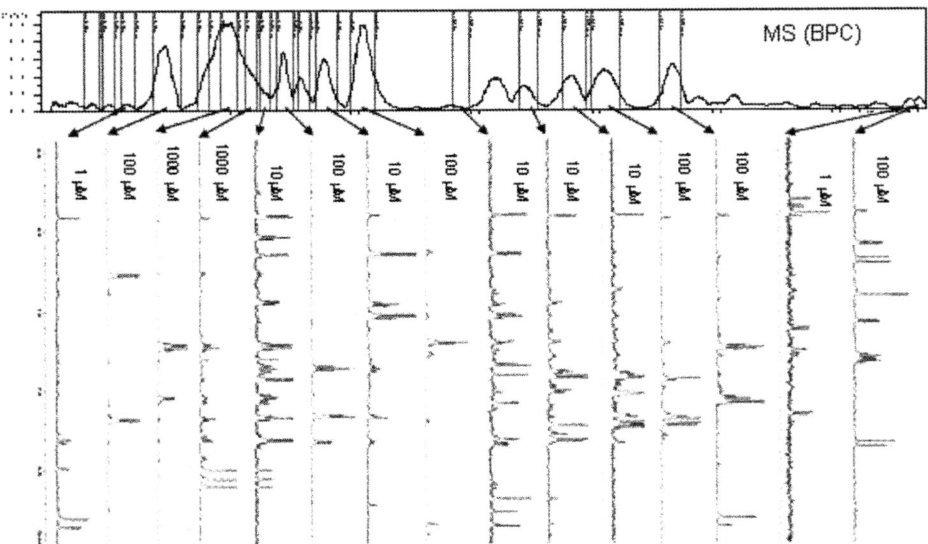

Figure 6 *Chromatogram (MS trace) resulting from a 60 μl injection of SPE1 elute. Within the chromatogram sections are indicated that were trapped fivefold on SPE². Below the corresponding NMR spectra are indicated with an indicated of the order of magnitude of the concentration.*

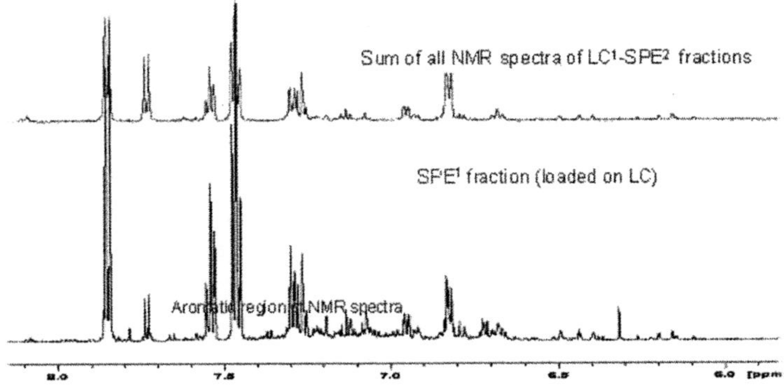

Figure 7 *Sum of all spectra produced by LC-SPE² (top) and spectrum of injected SPE¹ elute (bottom).*

4   CONCLUSION

In conclusion, an SPE^1-LC-MS-SPE^2-NMR identification strategy has been designed and refined for metabolites in complex biofluids. The approach has succesfully been applied to identify polyphenol metabolites produced by in vitro colonic microbial fermentation and in urine.

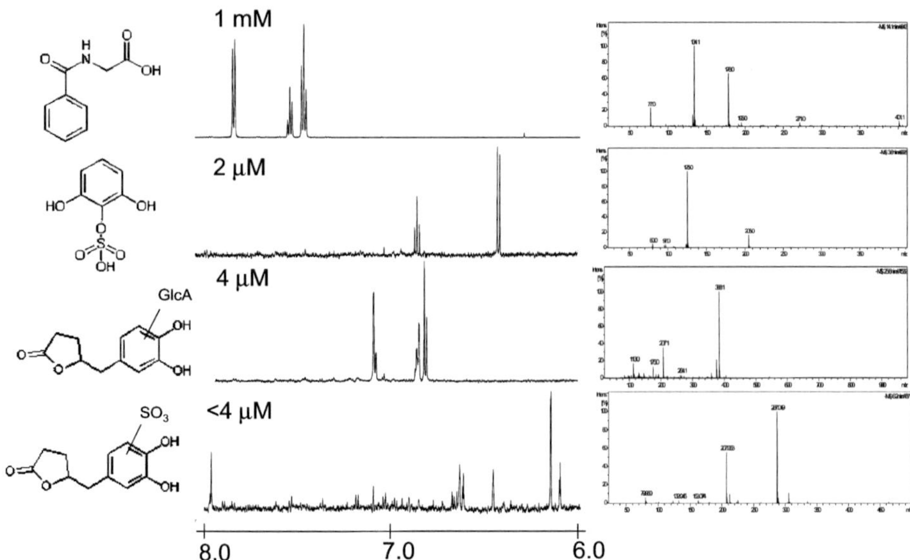

Figure 8 *NMR and MS spectra of identified conjugates of gut microbial polyphenol bioconversion products*

Acknowledgements

Part of this project was carried out within the research program of the Netherlands Metabolomics Centre, which is part of the Netherlands Genomics Initiative/Netherlands Organization for Scientific Research.

References

[1] G. Gross, D.M. Jacobs, S. Peters, S. Possemiers, J. van Duynhoven, E. Vaughan, T. Van de Wiele, *Journal of Agriculture and Food Science*, 2010, **58**, 10236-10246.

[2] J. van Duynhoven, E.E. Vaughan, D.M. Jacobs, R. Kemperman, E,J.J. van Velzen, G. Gross, L. Roger, S.Possemiers, A.K. Smilde, J. Doré, J.A. Westerhuis, and T. Van de Wiele, *PNAS*, doi/10.1073/pnas.1000098107

[3] C. Manach, J. Hubert, R. Lorach, A. Scalbert, The complex links between dietary phytochemicals and The complex links between dietary phytochemicals and human health deciphered by metabolomics. *Mol Nutr Food Res*, 2008, **53**, 1303-1315.

[4] A. Fardet, R. Llorach, A. Orsoni, J.-F. Martin, E. Pujos-Guillot, C. Lapierre, A. Scalbert, Metabolomics provide new insight on the metabolism of dietary phytochemicals in rats. *J Nutr* 2008, **138**, 1282-1287.

[5] C. Manach, G. Williamson, C. Morand, A. Scalbert, C. Rémésy, Bioavailability and bioefficacy of polyphenols in humans. I. Review of 97 bioavailability studies. *Am J Clin Nutr, 2005,* **81**, 230S-242S.

[6] J.K. Nicholson J.C. Lindon, *Nature* 2008, **455**, 1054-1056.

[7] C.H. Grun *et al.*, GC-MS methods for metabolic profiling of microbial fermentation products of dietary polyphenols in human and in vitro intervention studies. *J Chrom B*, 2008, **871**, 212-219.

[8]   Tang H, XiaoC., Wang Y. Important roles of the hyphenated HPLC-DAD-MS-SPE-NMR technique in metabonomics. *Magnetic Resonance in Chemistry,* 2009. S157-S162.

[9]   Nyberg NT, Baumann H, Kenne L. Application of solid-phase extraction coupled to an NMR flow-probe in the analysis of HPLC fractions. *Magnetic Resonance in Chemistry* 2001, **39(5)**, 236-240.

[10]   Godejohann M, Tseng LH, Braumann U, Fuchser J, Spraul M. Characterization of a paracetamol metabolite using on-line LC-SPE-NMR-MS and a cryogenic NMR probe. *Journal of Chromatography A* 2004, **1058(1-2)**, 191-196.

[11]   Sandvos M, Bardsley B, Beck TB, Lee-Smith E, North SE, Moore PJ, Edwards AJ, Smith RJ. HPLC-SPE-NMR in pharmaceutical development: capabilities and applications. 43 ed. 2005. 762-770.

[12]   A. Savage, G. Tucker, F. Wulfert, J. Van Duynhoven, C. Daykin, Nutrimetabolomics: development of a bio-identification toolbox to determine the bioactive compounds in grape juice, *Bioanalysis* 2010, **9**, 22-29.

[13]   L. Spiessser, D.M. Jacobs, in preparation

[14]   G.K. Pierens, A.R. Carroll, R.A. Davis, M.E. Palframan and R.J. Quinn Determination of Analyte Concentration Using the Residual Solvent Resonance in 1H NMR Spectroscopy, *Journal of Natural Products* 2008, **71(5)**, 810-813.

Imaging

RECENT CONCEPTS IN MRI

Franciszek Hennel, Stephanie Ohrel, Peter Ullmann, Markus Weiger, Robert Winkler.

Bruker BioSpin MRI, Ettlingen, Germany

1. CLASSIC MRI: LIMITATIONS

The methodology of Magnetic Resonance Imaging currently used on all commercially available machines is not much different from what was established two decades ago after the initial experiments with NMR localisation. It is based on a few solid fundaments that gave the technique mathematical simplicity, one of the reasons of its tremendous success.

Homogeneous excitation field. The transverse nuclear magnetisation, source of the NMR signal that allows the reconstruction of object's image, is produced (tilted from the longitudinal equilibrium direction) by a radio-frequency (RF) coil designed in such a manner that it produces a uniform RF field (B1) in the entire object. In this way, the same tilt angle, and thus the same signal and contrast can be observed everywhere. A deviation from B1 homogeneity is considered a problem. It becomes a serious trouble when object sizes become comparable with the length of the electromagnetic wave, a situation readily encountered in human MRI at 7T or higher fields.

Localisation with field gradients. The ingenious idea of the inventors of MRI was to link the NMR spectrum with the object's image by linearly mapping resonance frequencies to space with a gradient of the magnetic field (1,2). The general formula for the resonance signal evolving under time-and space-dependent magnetic field:

$$s(t) = \sum_r \rho(\mathbf{r}) \exp\left(i\gamma \int_0^t B(\mathbf{r},t')\,dt' \right), \qquad [1]$$

with $\rho(\mathbf{r})$ denoting the signal density at discrete location $\mathbf{r}$, becomes a familiar and easy-to-invert Fourier transformation:

$$s(\mathbf{k}) = \sum_r \rho(\mathbf{r})\,e^{i\mathbf{k}\cdot\mathbf{r}} \qquad [2]$$

provided the field depends on position via a linear gradient, i.e., $B = \mathbf{r}\cdot\mathbf{G}(t)$. The "wave vector" $\mathbf{k}$, time-integral of $\gamma\mathbf{G}(t)/2\pi$, describes the spatial modulation of nuclear magnetisation produced by the linear gradient. An MRI experiment consists of "navigating" through the k-space using field gradients as speed control, and processing the data collected during the trip with the inverse FT. Any deviation from linearity of the gradient leads to wrong assignment of signal sources in space and thus to image deformations. Clearly, such effects have always been unwanted in the classic MRI strategy.

Nyquist limit. The choice to reconstruct the image by Fourier-transforming the NMR signal (allowed through the use of field gradient) has consequences for the resolution and unambiguous field of view (FOV) of the image. These consequences are well known from the sampling theory: image elements become non-distinguishable if their spacing is smaller than the inverse of k-space range (uncertainty principle), while the FOV is given by the inverse of k-space sampling density (Nyquist limit) (3). Under these relations, increasing the resolution requires more data samples (to cover higher k-space range at constant spacing) and costs time. MRI is indeed a relatively slow technique, in many situations taking more time to collect the data then barely necessary for a sufficient signal-to-noise ratio. The Nyquist limit can be relaxed to some extent (in practice by factor of two to four) if an array of local RF coils is used in parallel for signal detection (4). Spatial sensitivity profiles of array elements represent an additional information source that allows recovering the image from undersampled data. But still, the question arises: does information content of an MRI image correspond to the amount of data dictated by the standard sampling principles? Or, have we got some prior knowledge about the image, meaning it contains less information we think, and it is possible to reduce the sampling beyond the Nyquist limit?

Echo acquisition. When the Fourier transform is applied to get the spectrum of the free induction decay (FID), the basic form of NMR signal, the result must be "phased" to separate the narrow real part from the broad imaginary "dispersion" component. Mathematically, the latter reflects the fact that the FID is there only on the positive side of the time axis. A spin echo, on the other hand, can be sampled symmetrically on both sides of the "time origin". The dispersion component disappears in that case and the tricky phasing operation can be replaced by the brute-force calculation of modulus spectra. Mainly for this pragmatic reason MRI has almost exclusively used echo signals (real reasons can be given in some cases when the actual phase of the spectra is of interest, e.g. in velocity mapping). Since the MRI signal is sampled in k-space rather than time, generation of echoes requires the trajectory to pass through k=0, which can be done not only with 180-degree RF pulses, but also with an inversion of gradient polarity. In any case, it requires some time, the so called Echo Time (TE), for the trajectory go away from the origin and back to form the required echo. Signals from many interesting sources, such as bones, tendons, teeth or lung, just to mention the clinical examples, have been considered MRI-invisible because their effective transverse relaxation times are shorter than practically achievable TE's. And all that just for the ease or image reconstruction? No wonder the idea of echo acquisition for MRI has been revisited...

2. PARALLEL TRANSMISSION

MRI usually requires spatial localisation, or restriction of a part of the object producing the signal to a volume fitting within the FOV. Otherwise, signals originating outside would be wrongly interpreted and cause the so called aliasing artefacts. In most acquisition methods, the active volume is limited to a thin slice requiring only a two-dimensional sampling trajectory and largely reducing the acquisition time. The classic method of MRI volume selection is based on narrow-band RF pulses applied in the presence of a field gradient, and thus fulfilling the resonance condition only for a given slice. Combining orthogonally selective pulses in echo sequences allows definition of 2- or 3-dimensional regions. An interesting alternative for such lengthy sequences has been proposed long time ago, but remains "asleep" for reasons explained further: so called multi-dimensional pulses, applied along with time-dependent gradients, doing the same job without necessity of echoes (5). The idea is based on the assumption, valid for low flip angles, that each short element of an RF pulse, P_t, produces its own signal starting with an amplitude proportional P_t and evolving without influence of later pulse elements. To get the signal produced at the end for a certain position r, one sums the components up:

$$S(\mathbf{r}) = \sum_t P_t \exp\left(i\gamma \int_t^{t_{end}} \mathbf{r} \cdot \mathbf{G}(t')\, dt \right) = \sum_t P_t\, e^{-i\mathbf{r}\cdot\mathbf{k}(t)} \qquad [3]$$

The k-space is introduced here again, this time to describe the modulation of transverse magnetisation by the time-dependent gradient during the pulse, with k=0 assigned to the end of the pulse. We recognize the Fourier-like relation between the k-space representation of the pulse shape and the spatial distribution of the initial signal, or the excitation pattern.

In practice, to design a pulse for a particular excitation pattern, one defines the pattern values on a discrete spatial grid, collects all these values in a vector S (even when the grid is 2- or 3-dimensional), and derives the discrete elements of the RF pulse by inverting the matrix form of the equation [3]:

$$S_r = \sum_t A_{rt} P_t \qquad [4]$$

where the design matrix A is composed of Fourier coefficients calculated along the k-space trajectory:

$$A_{rt} = e^{-i\mathbf{r}\cdot\mathbf{k}(t)} \qquad [5]$$

The success of the inversion, which is typically done in the least-squares sense, i.e. by applying pseudo-inverse of A to S, depends on the fidelity with which the pattern vector can be composed of the columns of A. To increase the fidelity for "demanding" patterns (such as a small, sharply defined volume in large object), more k-space points have to be traversed by the trajectory (to increase A's column space), meaning that the RF pulse has to be longer. On the other hand, long pulses become sensitive to short transverse relaxation times and resonance frequency offsets, leading, e.g., to fat being excited where it should not.

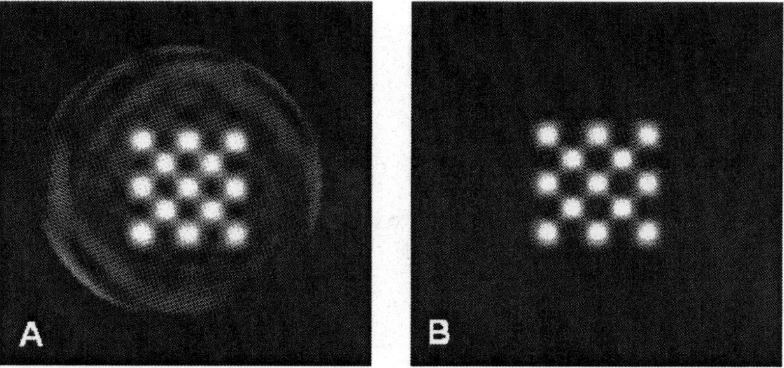

Figure 1: *Checkerboard pattern excited with a 2D spiral excitation pulse. Due to acceleration by increasing spiral pitch, artifacts appear when the pulse is transmitted by a volume coil (A). The same trajectory used for SENSE-transmission with a 3-element array gives a clean pattern (B). Acquired with Bruker BioSpec 9.4T scanner.*

A possibility to improve the excitation pattern fidelity at given pulse length, or to reduce the pulse length without trade-off on fidelity has been discovered due to the analogy of multidimensional excitation and imaging (6). With the latter, as already mentioned, sensitivity profiles of the receive-array used for parallel detection serve as additional means of spatial encoding. Excitation profiles can be improved in a similar way: pulses are applied in parallel through elements of a transmit-array (7). Still using the linear response regime (low flip angles), the excited pattern is a summed action of all coils:

$$S_\mathrm{r} = \sum_{n=1}^{\mathrm{n.coils}} \sum_t P_{nt}\, C_n(\mathbf{r})\, e^{-i\mathbf{r}\cdot\mathbf{k}(t)} \qquad [6]$$

where $C_n(\mathbf{r})$ denotes the complex sensitivity of coil n at position r. Despite the double summation, each pixel of the excitation pattern is still given by a linear combination of pulse elements transmitted by all coils (P_{nt} is what coil n transmits at time-point t). If all these elements are placed in a single long vector P, the equation takes the familiar form of [4], with the design matrix having $N_\mathrm{coils} \times N_\mathrm{time\ points}$ columns. If the sensitivities of the coils differ enough, the additional columns increase the column space of the design matrix and thus the chance to obtain the required pattern more reliably. Figure 1 presents images of a 2-dimensional checkerboard pattern selected inside a water filled phantom with a spiral-trajectory pulse. Transmitted with a volume resonator, the pulse does not do its job properly: elements outside of the pattern are excited due to insufficient density of the spiral trajectory. When a 3-element array is used to transmit 3 different, properly calculated pulses, these artefacts disappear.

A critical issue related to parallel-transmission pulse design is the measurement of the coil transmission profiles, or B1-field maps. The quest for the best (sensitive, high-range and fast) B1-mapping method is not yet concluded. One of the most recent proposals is based on the Bloch-Siegert shift (8).

Parallel transmission will probably allow extending the selective excitation to three dimensions, an idea known for quite some time, but regarded as unpractical. Recently, interleaved multi-volume excitation has been demonstrated using concentric shell-trajectories with pulse durations of less than 5 milliseconds (9).

3. HIGH ORDER FIELD ENCODING

MRI images often contain huge amount of superfluous information: the entire field of view is depicted with the same resolution as the actual regions of interest. The straightforward way to reduce the information contents by zooming the FOV to the region of interest is hindered by the aliasing effect, which can be reduced by selective saturation or excitation in some cases, but never removed completely. The alternative is to keep the FOV bigger than the object and "zoom" the resolution only. Put differently, to get rid of the shift-independent resolution of standard MRI.

Shift-independent resolution results from the Fourier relation between data and image domains, which, in turn, is due to the use of linear field gradients for encoding. However, when we look at Eq. [1], solutions for the image ρ_r can be envisaged for other field-space dependencies than linear. Written for a discrete grid of image elements and time points, this equation takes again the linear form:

$$S_t = \sum_r A_{tr} \, \rho_r \tag{7}$$

with the design matrix

$$A_{tr} = \exp\left(i\gamma \int_0^t B(t',\mathbf{r})dt' \right) \tag{8}$$

and can be solved for the image ρ_r by least-squares fit. The resolution will now depend on the position in space. It can be calculated by observing what happens to a point signal source upon the reconstruction, the effect described by the matrix form of the so called spatial response function (10):

$$\mathbf{SRF} = \mathbf{A'} \; \mathbf{A} \tag{9}$$

where A' is the reconstruction matrix (pseudoinverse of A). Element $SRF_{r'r}$ tells the image intensity at position r' provided the source was at r. SRF is thus diagonal in the ideal case. However, the "diagonal" gets broader (resolution is lost) when design matrix columns and reconstruction matrix rows belonging to different positions are correlated. Whether such correlations are avoided depends on the ability of $B(t,\mathbf{r})$ to generate different modulations for neighbouring positions in possibly short time. Put differently, local resolution depends on the local gradient of the encoding field.

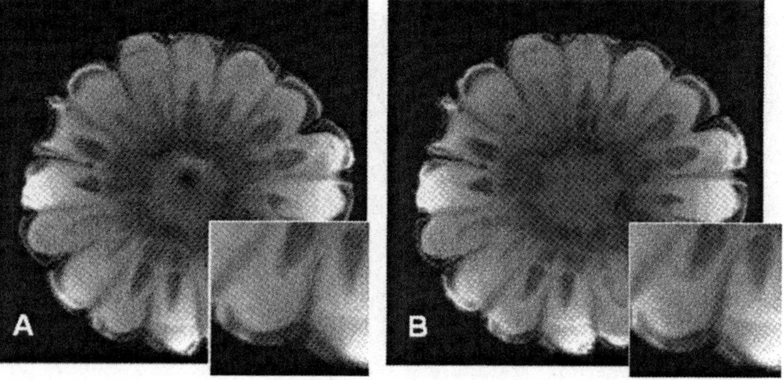

Figure 2: *Images of a boiled corn cob obtained with PATLOC encoding fields (A) and with standard gradients (B) in the same time. Resolution is locally increased in the external parts with the PATLOC technique. Acquired with Bruker BioSpec 9.4T scanner.*

In a recently described PATLOC technique (parallel acquisition technique using local gradients) (11) the encoding field is generated by a polar arrangement of local coils, which produce maximal gradients in the external part of the object while keeping it zero in the centre. Realised experiments used an arrangement of four coils combined in two pairs to produce fields similar to the 2xy and $x^2\text{-}y^2$ shims. The reconstruction procedure was based on the Fourier transform, which can be re-introduced to the encoding equation by factorizing the field into a scalar product of a time dependent scaling vector (representing the coil currents) and a constant vector of spatial dependencies (B0 maps of the coils):

$$B(t,\mathbf{r}) = \mathbf{I}(t) \cdot \mathbf{\Phi}(\mathbf{r}) \tag{10}$$

This allows definition of a k-space vector (as the integral of **I**) and calculation of an image in distorted coordinates (those of ϕ) with inverse-FT, before un-warping it back to the actual space. This approach makes it even clearer that the resolution of the final image is position-dependent via local $\partial\Phi_i / \partial r_j$ gradients.

Fields produced by PACTLOC coils encoding are symmetric, meaning that only symmetric images (representing a folded object) result from the initial reconstruction. A disambiguation has to be performed using additional localization means, namely, the sensitivity profiles of the receive-array. PATLOC is therefore inherently a parallel acquisition method.

For a given duration of the experiment, PATLOC will achieve a higher resolution in the external part of the object than standard, gradient-based MRI. This is advantageous in several situations, e.g. when cortical regions of the brain are of interest. Further advantages are the reduction of field of view, achievable also for linear coil arrangements, and lower peak field values generated by the coil. The latter increases the safety range within which a sequence can run without causing peripheral nerve stimulation.

4. UNDERSAMPLED RECONSTRUCTION

Except for the rare cases of automatic assessment, MRI images typically end up in front of human eyes. These (along with the rest of our visual system) appear to be quite tolerant to certain manipulations of the image, even those strongly reducing the information contents (and file size), as the tremendous success of photo compression demonstrates. Differently speaking, images are compressible, which means, they contain less useful information than can be contained in all pixels. One is obviously tempted to use the compressibility feature to reduce the amount of data needed for image reconstruction and thus the measurement time. This strategy has been explored in the recent years under the appellation of "compressed sensing" (12).

Mathematically, an image is compressible when it can be linearly transformed to a sparse matrix, i.e., a matrix whose most elements are zero (or can be replaced by zeros without alarming human vision). Some MRI images are sparse themselves, e.g. angiograms. Most images can be "sparsified" by transformations based on local frequency decomposition such as wavelet transforms.

The idea of compressed sensing consists of searching of an image that not only matches the undersampled data (an infinite number of images do that), but at the same time limits the number of non-zero components, or the size of the support of the sparse transform. Technically speaking, one tries to minimize the L0 norm of the transform under the constraint of the measured data.

One of the conditions of the success of this reconstruction strategy is that the undersampling does not lead to coherent artefacts in the sparse transform domain, but instead, produces a noise-like contribution. This allows searching for the image by taking the strongest point in the sparse transform derived from the incomplete data, adding this point to the support of the true transform, subtracting the contribution of that support from the signal, selecting the next point in the transform of the resulting signal, and so on — a procedure known as "orthogonal matching pursuit". Numerous other techniques exist that allow including additional *a priori* information about image properties, such as smoothness, in the optimization criteria.

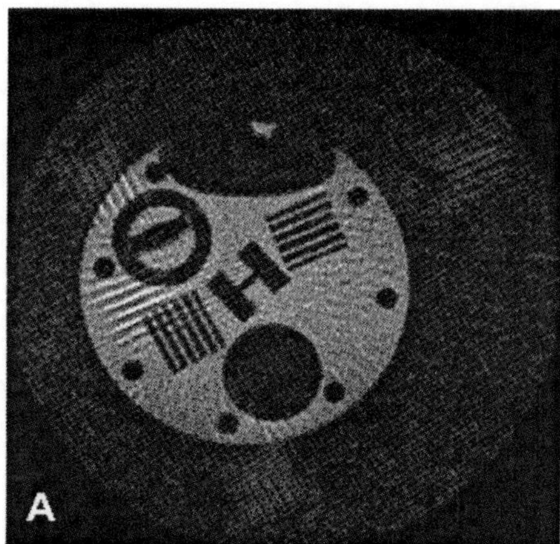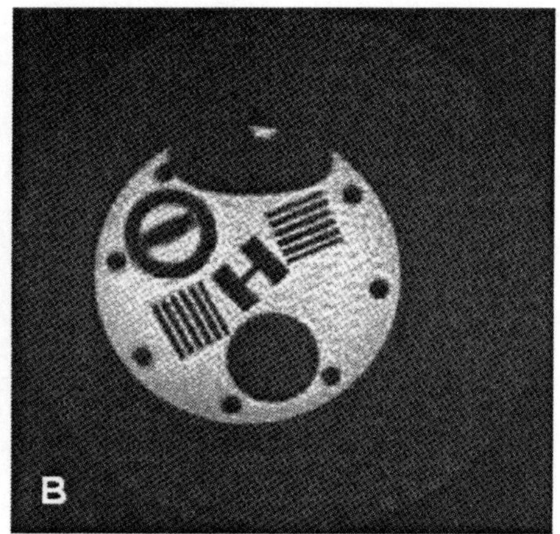

Figure 3: *Radial scan of a test object reduced to 1/8 of the nominal number of projections. Streaking artefacts that appear in the standard zero-filled FT (A) are avoided with compressed sensing reconstruction (B). Acquired with Bruker PharmaScan 7T scanner.*

Computing time still seems to be the factor hindering the wide use of compressed sensing, but the field of applications is growing. To name a recent one, CS appears useful in experiments with hyperpolarized nuclei (13) where, due to the un-recoverability of the magnetization, less encoding steps means not only faster measurement, but higher signal as well.

5. NON-ECHO ACQUISITION

The alternative to echo-based MRI is the direct spatial encoding of the free induction decay signal produced with a single RF pulse, which implies that the sampling trajectory starts in the centre of k-space. In the simplest case, when the signal evolves in the presence of a constant gradient, the sampling pattern consists of a set of radii starting at k=0 and reaching the k-space range given by desired image resolution, each radius corresponding to a new excitation and to another direction of the readout gradient. The total number of excitations may be reduced if the trajectories are bent in a spiral manner. However, the straight radial pattern provides the minimum readout duration at a given gradient strength and thus the minimum resolution loss due to T2 decay. Radial sampling of FID signals is actually known since the pioneering work of Lauterbur and owes the come-back of its popularity to the interest in short-T2 imaging.

The critical element of the FID-based MRI of short-T2 objects is the excitation, which should ideally take place while the readout gradient is turned on. This is challenging for two reasons: the excitation must cover a broad frequency band (equal to that spanned by the object in the presence of the readout gradient), and, the reconstruction must deal with the fact that the first sample of the signal is taken only after a delay needed to switch the spectrometer hardware from the transmission to reception mode. An easy solution to both problems has been proposed with the UTE (ultra-short TE) technique (14), where the readout gradient is ramped up after the excitation. This unavoidably leads to a delay of the central k-space sample and introduces a minimum "echo time" of the order of a few microseconds. A truly "zero-TE" acquisition must use a broadband excitation during the readout gradient, and reconstruct the image correctly despite the missing central k-space portion.

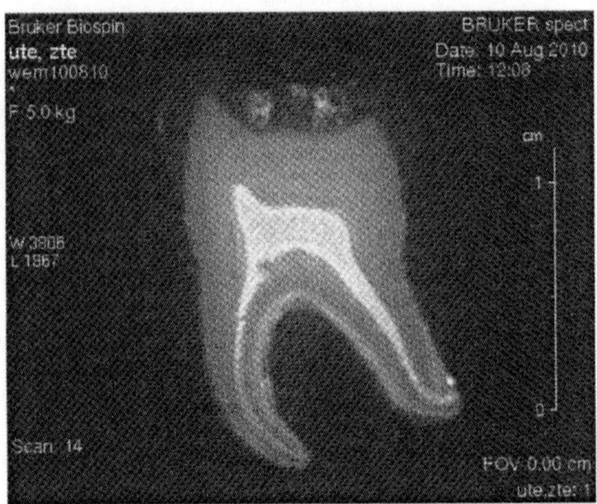

Figure 4: *MRI of an extracted human tooth obtained with the zero echo time (ZTE) sequence based on hard pulse excitation in a 11.7 T Bruker micro-MRI system. The excellent visibility of dental layers including enamel (dark grey) is made possible by the sensitivity of ZTE to T2 values below 200 microseconds.*

The task of exciting the broad frequency range occupied by the object can be achieved in a straightforward manner by applying a short amplitude modulated (hard) RF pulse, of the order of 1-5 microseconds. Pulses that short sometimes do not allow reaching the optimum flip angles. Alternatively, long frequency-sweep pulses can be used, as proposed in the SWIFT method (15). In this case, the reconstruction requires the signal to be sampled already during the RF pulse, which, for practical reasons, can only be performed during short gaps in the RF transmission. In both versions of Zero-TE MRI, one deals with signals which are not continuously sampled. A short initial part of the signal in the hard-pulsed version, or periodic signal gaps in sweep-MRI are missing. Such acquisition schemes can be conveniently processed with algebraic matrix inversion instead of inverse Fourier transform (16), which treats unknown signal portions as zero and leads to baseline artefacts. The linear relation between the time domain signal and the object density, in the form of Eq. [7], can be obtained by adding a sum over spatial positions to the equation [3], which describes the signal produced by an RF pulse. The encoding matrix is similar, and contains an additional sum over RF pulse elements P_n:

$$A_{tr} = \sum_n P_n \exp\left(i\gamma \int_{t_n}^{t} B(t',\mathbf{r})dt' \right) \qquad [11]$$

This approach allows treatment of signals acquired during an RF pulse, in gaps during pulse transmission and after the pulse as well.

The hard-pulse and sweep-pulse versions of zero-TE imaging have been compared in a recent study with respect to their ability of depicting short-T2 signals and to their SNR-efficiency (17). It appears that, while the resolution dependence on T2 is identical for both, the SNR efficiency is higher in the hard-pulse version for RF intensities usually available on animal- and micro-MRI systems. This is due to the fact that the periodic detection masking needed in the sweep approach deteriorates the SNR to an extent that is higher than the losses caused by sub-optimal flip angles of the hard pulses.

6. SUMMARY

The number of new ideas that emerged in the domain of MRI technology in the recent years, only few of them arbitrarily mentioned in this review, lets us believe that this imaging modality has still not reached it ultimate shape. Due to increasing computing power, reconstruction methods long considered unpractical, such as the direct fitting of the image instead of the estimating it with Fourier transform become attractive and allow image recovery from incomplete datasets. This leads not only to faster scans, as in it is the case of compressed sensing, but also opens new image contrast possibilities, such as the zero-TE imaging. Interest in alternative tools for spatial encoding of the NMR signal, such as PATLOC gradients or spatially-modelled profiles of parallel RF-transmission arrays will surely result in a further increase of MRI speed and efficiency. Observing research activities in this fascinating field will be certainly worthwhile in the nearest future.

References

1. P. Lauterbur, *Nobel Lecture*, 2003, available in internet from nobelprize.org
2. P. Mansfield, *Nobel Lecture*, 2003, available in internet from nobelprize.org
3. R.N. Bracewell The *Fourier transform and its applications* 2nd ed., chapter 10, McGraw-Hill, New York, 1986
4. Pruessmann KP, Weiger M, Scheidegger MB, Boesiger P. *Magn Reson Med.* 1999 **42**, 952-62.
5. J. Pauly, D. Nishimura and A. Macovski, *J. Magn. Reson.* 1989, **81**, 43-56
6. Börnert P, Aldefeld B., *MAGMA*, 1998, **7**,166-78.
7. Katscher U, Börnert P, Leussler C, van den Brink JS., *Magn Reson Med.* 2003 **49**, 144-50.
8. Sacolick LI, Wiesinger F, Hancu I, Vogel MW. *Magn Reson Med.* 2010, **63**,1315-22.
9. Schneider JT., Kalayciyan R., Haas M., RuhmW., Doessel O., Hennig J., Ullmann P. *Proc. ISMRM* 2010, Stockholm, p. 103.
10. Pruessmann KP, *NMR Biomed.* 2006, **19**, 288-99.
11. Schultz G, Ullmann P, Lehr H, Welz AM, Hennig J, Zaitsev M, *Magn Reson Med.* 2010 **64**,1390-403
12. Lustig M, Donoho D, Pauly JM, *Magn Reson Med.* 2007, **58**,1182-95.
13. Ajraoui S, Lee KJ, Deppe MH, Parnell SR, Parra-Robles J, Wild JM, *Magn Reson Med.* 2010 **63**,1059-69.
14. Glover GH, Pauly JM, Bradshaw KM, *J Magn Reson Imaging.* 1992, **2**, 47-52
15. Idiyatullin D, Corum C, Park JY, Garwood M, *J Magn Reson.* 2006, **181**, 342-9.
16. Weiger M, Hennel F, Pruessmann KP, *Magn Reson Med.* 2010, **64**,1685-95.
17. Weiger M. Pruessmann KP, Hennel F, *Magn. Reson Med.* 2011, in press.

DIFFUSION-WEIGHTED NMR MICRO-IMAGING OF LIPIDS: APPLICATION TO
FOOD PRODUCTS

Sylvie Clerjon*, Jean-Marie Bonny

STIM, UR 370 QuaPA, Centre INRA de Theix, 63122 St-Genès-Champanelle, France
**Tel. 33(0)4 73 62 45 93, Fax. 33(0)4 73 40 89, Email sylvie.clerjon@clermont.inra.fr*

1. INTRODUCTION

MRI has no equivalent for the non-destructive analysis of fat distribution and content at
voxel scale. For fat distribution assessment, many applications take advantage of the
clearly marked contrast between fat and surrounding tissues using standard T_1, T_2 or proton
density-weighted sequences (Foster, Hutchison, Mallard & Fuller, 1984; Groeneveld,
Henning & Kallweit, 1988). Contrast can be improved using fat- or water- suppression
techniques that can be applied using inversion preparation schemes (Bydder & Young,
1985; Laurent, Bonny & Renou, 2000) and Dixon spectroscopic imaging (Ma, 2008).
Other authors have published comparisons between these NMR fat imaging techniques
(Bernard, Liney, Manton, Turnbull & Langton, 2008; Papan, Boulat, Velan, Fraser &
Jacobs, 2007).

In food science, intramuscular fat influences some important meat quality
characteristics e.g. marbling, sensory properties and keeping qualities, and also diffusion of
water and salt in cured/smoked products. Fat distribution through the muscle is an
important factor for meat palatability, juiciness and the taste and smell of final products.
Thus Tingle et al. (1995) describe T_1- weighted and STIR imaging of fat in lamb
longissimus dorsi muscle. Fat entering food products during cooking must also be
monitored to improve their nutritional properties by reducing fat penetration. For instance,
oil penetration in Japanese tempura after frying is quantified by T_{1w}, T_{2w} and proton-
density magnetic resonance imaging (Horigane, Motoi, Irie & Yoshida, 2003). Other food
applications such as cheese ripening (Mahdjoub, Molegnana, Seurin & Briguet, 2003) or
lipid migration in chocolate (Walter & Cornillon, 2002) require MRI of fat for process
optimization.

Here we present an original approach for imaging fat distribution in muscle based on
diffusion-weighted imaging. This technique takes advantage of the large difference
between the apparent diffusion coefficients (ADC) of protons in nuclei belonging to water
molecules and those in fat. As the dynamics of Brownian motion is related to molecular
size, a large difference in ADC is generally observed in biological tissues and food
matrices, water diffusing much more rapidly than fat. Assuming Gaussian diffusion
behavior, the image magnitude obtained using a diffusion-weighted sequence is attenuated
exponentially i.e. by a factor $\exp^{-b.\mathrm{ADC}}$, where b is the extrinsic diffusion weighting of the
sequence. The more mobile the protons are, the more their corresponding signal is
attenuated. Using a sufficiently large b value, it is possible to attenuate (and even suppress)
the signal of water so that only that of fat becomes detectable.

The diffusion contrast between water and fat has been pointed out in various
biomedical contexts and particularly in muscle. As it is closely linked to our present aim,
we present examples of these studies below. Although Ababneh et al (2005) in their paper
describe the ability of diffusion imaging to highlight water in muscle edema, they also

show diffusion-weighted images acquired at high b values of 3000 s.mm^{-2} and explain that in these very high diffusion weighted images, the signal from slowly-diffusing lipids in bone marrow is prominent, as recently discussed by Lehnert et al. (2004). Diffusion-weighted imaging has also already been used in musculoskeletal systems for fat infiltration control in myopathies (Qi, Olsen, Price, Winston & Park, 2008) and evaluation of myoskeletal tissue (Karatopis, Anastasopoulos, Drakopoulos, Douskou, Panagiotakis & Kandarakis, 2008).

For *in vivo* applications, the main difficulties come from artifacts caused by patient motion and by susceptibility differences near interfaces when echo-planar diffusion-weighted techniques are used. Both difficulties may be circumvented in food science because most structural studies are done on a static food matrix using CPMG-based sequences not prone to susceptibility artifacts. Here we present two examples of high resolution fat DWI in muscle food. The first is designed to image intrinsic intramuscular fat content in raw beef and the second to image oil penetration in pork after plane frying. A procedure for selecting a suitable b value is presented.

2. MATERIALS AND METHODS

Optimization of b value for DWI of fat in muscle

A large number of sequences have been proposed for generating images of fat distribution that exploit the differences in T_1 (e.g. inversion recovery) and (or) resonance frequencies between fat and water (e.g. CHESS or Dixon phase-sensitive technique) (Laurent, Bonny & Renou, 2000).

These methods are based on different contrasts; the higher these intrinsic contrasts are, the more robust is the method. In muscle, which is relevant here, T_1 contrast between muscle fibers (mainly composed of water) and fat is given by:

$$\frac{T_{1(W)} - T_{1(F)}}{T_{1(W)}} \approx 3\ \% \tag{1}$$

where $T_{1(W)}$ is the longitudinal relaxation time of water and so of muscle fibers and $T_{1(F)}$ is the longitudinal relaxation time of fat. This contrast also depends on the static magnetic field and the temperature, which leads to practical difficulties and variability in measurements. Chemical shift contrast between water and fat is 3.5 ppm. This contrast is poor compared with the linewidth of the proton spectrum, especially at low field. Also, this frequency difference is sensitive to the spatial variation of magnetic field strength in the sample, which becomes a major problem in any frequency-selective fat imaging technique.

Following Henkelman, Stanisz, Kim & Bronskill (1994), we checked that muscle fibers displayed bi-exponential diffusion behavior: a bulk component with a fast $ADC_W \sim 1.2 \times 10^{-3} mm^2/s$ and a minor one for which $ADC \sim 0.07 \times 10^{-3}\ mm^2/s$ at 20°C. ADC_F of lipids is equal to $\sim 0.01-0.02 \times 10^{-3} mm^2/s$ at 20°C (Bonny & Renou, 2004; Bonny, Sante-Lhoutellier and Renou, 2005). These values are consistent with those reported by Lehnert et al. (2004) in subcutaneous fat of human subjects. In a first approximation, neglecting the minor diffusion component of water in muscle, the diffusion contrast is equal to:

$$\frac{ADC_W - ADC_F}{ADC_W} \approx 100\ \% \tag{2}$$

This contrast is much higher than T_1 or chemical shift contrasts. Consequently, the signal of water obtained with a diffusion-weighted sequence is more quickly attenuated in the muscle fibers than in lipids.

Here our aim was to attenuate the signal of water sufficiently compared with that of fat. The signal-to-noise ratios of water and fat (respectively SNR_W and SNR_F), for a given voxel volume V, are:

$$SNR_W = \left(\frac{V}{V_0}\right).SNR_{0W}.\left[f.\exp(-b.ADC_f) + (1-f).\exp(-b.ADC_s)\right] \qquad (3)$$

$$SNR_F = \left(\frac{V}{V_0}\right).SNR_{0F}.\exp(-b.ADC_F) \qquad (4)$$

where SNR_{0W} and SNR_{0F} are the signal-to-noise ratios of reference at V_0 respectively for water and fat, and f is the fraction of fast-diffusing water.

To determine the optimized b value, we measured the DW signal of lean meat and fat for b values in the range 100–200,000 s.mm^{-2}. In our first application fat was intrinsic intramuscular fat; in our second application it was frying oil, i.e. olive oil. Preliminary experiments, in accordance with literature (Ababneh at al., 2009), having shown that intramuscular fat ADC is close to that of olive oil, Figure 1 presents the DW signal versus b values for lean meat and olive oil. For a given resolution, i.e. for a given V_0, SNR_0 depends only on sequence parameters (T_R, T_E, angles, etc.) and on experimental conditions (B_0, antenna response, proton density and temperature). To obtain SNR_{0W} and SNR_{0F}, we performed preliminary measurements at $b = 0$ and low resolution. These values are the signal intensity at $b = 0$ in Figure 1.

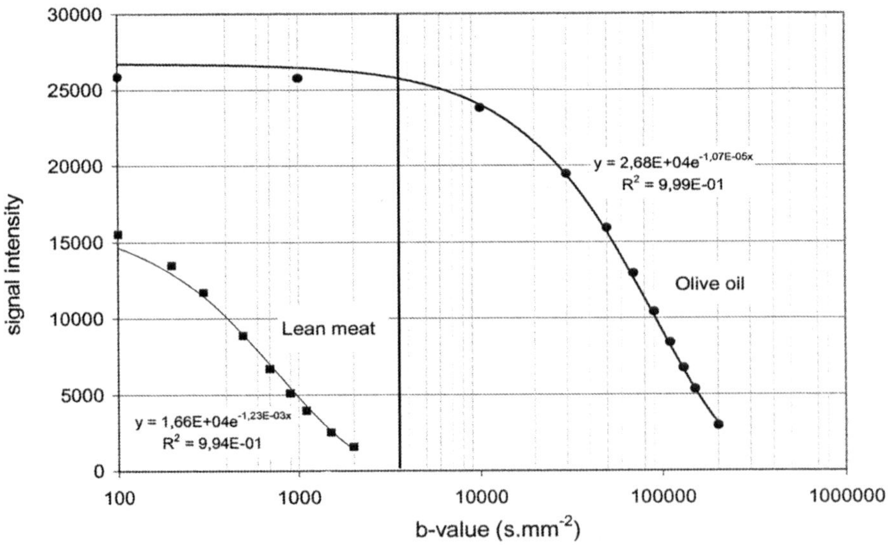

Figure 1. *Diffusion-weighted signal as a function of b value measured by imaging in lean meat and olive oil.*

We can see in Figure 1 that increasing b improves contrast between muscle fibers and fat up to muscle fiber saturation ($b \sim 2000$ s.mm^{-2}). For a better spatial resolution, i.e. for a smaller voxel volume, the b value required for attenuating the signal of muscle fibers decreases and leads to a higher signal intensity in fat (lower attenuation due to both diffusion and T_2 effects), and so to a higher contrast. This explains why diffusion-weighted imaging is especially well adapted to 3D micro-imaging. Finally, in the specific imaging conditions illustrated in Figure 1, the best compromise b value is 3500 s.mm^{-2}.

Against this general background, two applications of lipid imaging in muscle food products using diffusion-weighted contrast are presented in the following two sections.

Samples and MRI for intramuscular fat in ex vivo *beef muscle*

The samples studies were pieces of raw *infraspinatus* bovine muscle at ~ 20°C.

MRI experiments were performed at 200 MHz on a Biospec 47/40 Bruker (Bruker, GmbH, Ettlingen, Germany) with fast BGA12 gradient hardware and a 72 mm-diameter birdcage radiofrequency coil used for both emission and reception.

A pulsed gradient SE sequence was implemented that included a composite nonselective pulse to limit the refocusing errors due to B_1 inhomogeneities over the whole field of view.

For these samples, the suitable b value was calculated according to the previous scheme using f = 95 %, ADC$_f$ = 1.20×10^{-3} mm^2/s and ADC$_s$ = 0.07×10^{-3} mm^2/s. The desired b value was equal to 1200 s/mm^2 in the case of an isotropic acquisition with V = 0.064 mm^3. The latter corresponds to a field of view of 51.2 mm and a matrix size of 128, giving an acquisition time of 28 min (T_R = 100 ms).

Samples and MRI for oil penetration in pork meat after plane frying

This study was part of a project designed to improve the nutritional properties of meat by fatty acid supplementation in animal feed. This supplementation resulted in lipid profiles of meat rich in essential fatty acids. However, it was important to check whether conventional cooking by plane frying degraded the nutritional properties of the meat. For this purpose we quantified the penetration of oil during a cooking experiment by NMR micro-imaging.

Samples were cylindrical meat pieces of diameter 25 mm and length 50 mm. They were subjected to plane frying in olive oil for 6 min at 130°C at one end.

MRI experiments were performed at 400 MHz on an Avance DRX400 microimaging system (Bruker, GmbH, Ettlingen, Germany) with an actively shielded gradient coil. The samples were placed in a 32 mm-diameter birdcage radiofrequency coil used for both excitation and signal reception.

To optimize the b value of the diffusion-weighted SE sequence, we took olive oil and lean meat signal measurements at several b values (Figure 1) to calculate ADC values. For these preliminary experiments we used a pulsed gradient SE sequence with the following MRI parameters for olive oil: T_R/T_E = 5000/42.4 ms and NA = 8, and for lean meat: T_R/T_E = 5000/18.2 ms and NA = 10. We see in Figure 1 that a b value at 3500 s.mm^{-2} was the best compromise for water extinction and maximal olive oil signal.

Other preliminary experiments showed there was no significant difference in signal intensity, for a given b value, between olive oil and intramuscular fat. These preliminary results are consistent with the literature (Ababneh at al., 2009). Hence our MRI method cannot discriminate between intrinsic intramuscular fat and absorbed oil. Oil penetrating during frying must be quantified by comparison of images of raw and fried states for each sample.

Muscle fibers and olive oil ADC and SNR$_0$ being known, we adjusted all sequence parameters to obtain a strong enough signal at the resolution we needed to monitor oil penetration. Voxel volumes were $100 \times 500 \times 500$ µm^3, the higher resolution (100 µm) corresponding to the axial direction of the sample, i.e. the oil penetration direction. The other acquisition parameters were: T_R/T_E = 500/16.6 ms and NA = 8. Using this protocol, each acquisition took 4 h 35 min at a constant temperature of 25°C.

3. RESULTS

PGSE images of the raw bovine muscle sample are shown in Figure 2. Because of the high spatial resolution and the voxel isotropy, intramuscular fat distribution is depicted here with the same quality irrespective of incidence. For example, the continuity of the fat network along the muscle fascicles is clearly demonstrated.

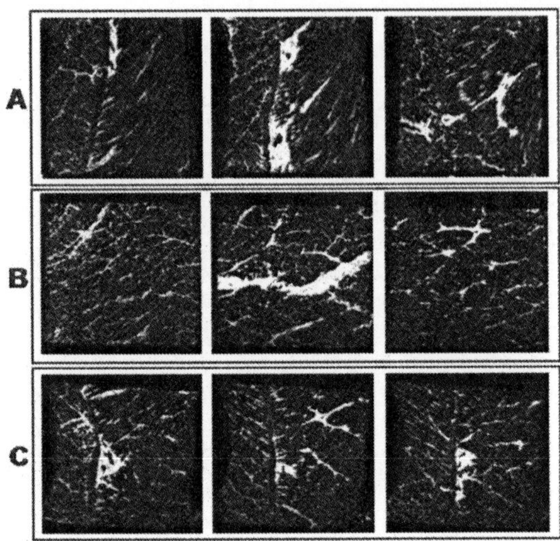

Figure 2. *Successive planes (from left to right, spacing of 12.8 mm) obtained from the 3D-PGSE block of the infraspinatus bovine muscle in three orthogonal incidences (A, B, C).*

Fat imaging was performed before and after frying to monitor the spatial profile of the oil penetration in pork during the frying process (Figure 3). For each acquisition, a profile of fat in the direction perpendicular to the cooking surface was estimated by image processing. The method corrects the non-flatness of the face after cooking and the intrinsic content of intramuscular fat in the sample. It is therefore possible to access average profiles of fat first on a raw sample, and then cooked, and the average profile of the oil penetration during the frying process (Figure 4). Diffusion-weighted NMR micro-imaging associated with image analysis allows the penetration of fat in a meat product to be monitored. Here we show penetration of oil to a depth of about 5 mm in meat during plane frying at 180°C.

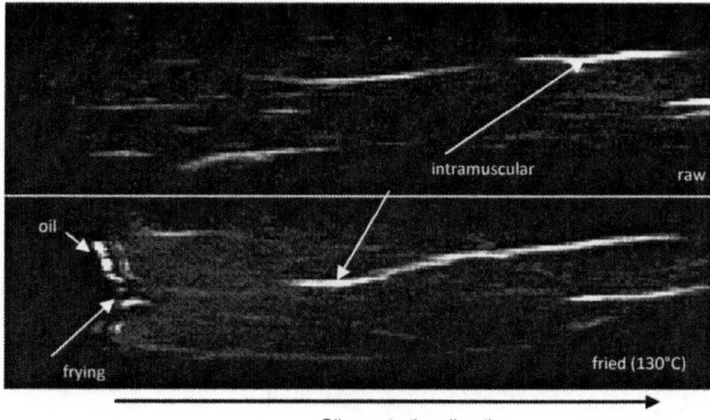

Figure 3. *Fat signal before and after frying. Hypersignal corresponds to NMR fat signal.*

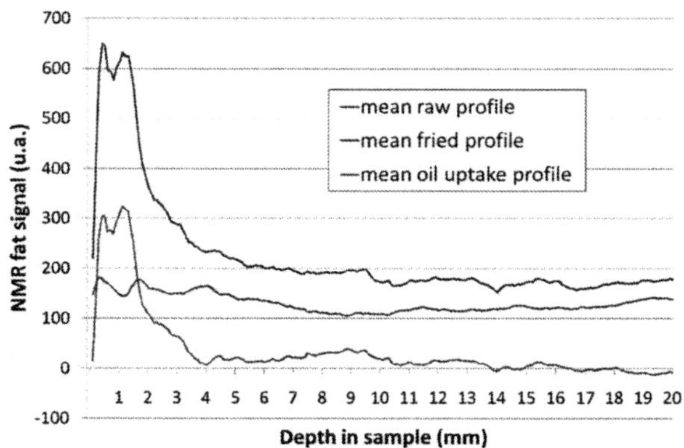

Figure 4.*Average profiles of fat in a raw sample and in a 180°C fried sample and the average profile of the oil penetration during the frying process.*

4. DISCUSSION

Diffusion-weighted imaging is especially well-suited to a high-resolution depiction of fat distribution as the *b* value needed to attenuate the myofibre signal decreases with the voxel volume V. This method is based on the high contrast between water and fat ADC, and these ADC do not depend on the static field, which is an advantage in low-field whole-body systems for biomedical applications.

The drawbacks of this method are the sensitivity of diffusion-weighted images to bulk motion, limiting its use to motionless samples. Also, any factor inducing a variation in the ADC contrast between fat and myofibers must be taken into account as it modifies settings (e.g. temperature and drying).

In conclusion, diffusion-weighted imaging is a simple approach to imaging the distribution of intrinsic intramuscular fat. DWI may be used in numerous medical applications where high resolution fat imaging is needed. It may also have useful applications in food science, since fat distribution influences some important meat quality characteristics. Finally, extrinsic lipids penetrating food products during processing can be controlled to improve nutritional qualities.

References

Z. Ababneh, H. Beloeil, C.B. Berde, G. Gambarota, S.E. Maier, R.V. Mulkern, Biexponential parameterization of diffusion and T-2 relaxation decay curves in a rat muscle edema model: Decay curve components and water compartments. *Magnetic Resonance in Medicine,* 2005, **54**, 524-531.

Z.Q. Ababneh, H. Beloeil, C.B. Berde, A.M. Ababneh, S.E. Maier, R.V. Mulkern, In vivo lipid diffusion coefficient measurements in rat bone marrow. *Magnetic Resonance Imaging*, 2009, **27**, 859-864.

P. Bernard, G.P. Liney, D.J. Manton, L.W. Turnbull, C.M. Langton, Comparison of fat quantification methods: A phantom study at 3.0 T. *Journal of Magnetic Resonance Imaging,* 2008, **27**, 192-197.

J.M. Bonny, J.P. Renou. Multiexponential modeling of diffusion in meat at large b-values by MRI. 7th International Conference on the Applications of Magnetic Resonance in Food Scienc, 2004, Copenhagen.

J.M. Bonny, V. Sante-Lhoutellier, J.P. Renou in: S. B. Engelsen, P. S. Belton, and H. J. Jakobsen Eds., *Magnetic Resonance in Food Science*, RSC, Cambridge, 2005, pp. 141-147.

G.M. Bydder, I.R. Young, Clinical use of the partial saturation and saturation recovery sequences in MR imaging, *J. Comput Assist Tomogr.*, 1985, **9**, 1020-32.

M.A. Foster, J.M.S. Hutchison, J.R. Mallard, M. Fuller, Nmr Pulse Sequence and Discrimination of High-Fat and Low-Fat Tissues. *Magnetic Resonance Imaging,* 1984, **2**, 187-192.

E. Groeneveld, M. Henning, E. Kallweit Growth patterns and carcass evaluation in pigs by NMR measurements. in: M. Henning, E. Kallweit, and E. Groeneveld Eds., *Application of NMR techniques on the body composition of live animals*, Elsevier Applied Science, New York, 1988, pp. 137-148.

R.M. Henkelman, G.J. Stanisz, J.K. Kim, M.J. Bronskill, Anisotropy of Nmr Properties of Tissues. *Magnetic Resonance in Medicine,* 1994, **32**, 592-601.

A.K. Horigane, H. Motoi, K. Irie, M. Yoshida, Observation of the structure, moisture distribution, and oil distribution in the coating of tempura by NMR micro imaging. *Journal of Food Science*, 2003, **68**, 2034-2039.

A. Karatopis, S. Anastasopoulos, S. Drakopoulos, M. Douskou, G.S. Panagiotakis, I. Kandarakis, Molecular Imaging of the myoskeletal system through Diffusion Weighted and Diffusion Tensor Imaging with Parallel Imaging Techniques. 2008 Ieee International Workshop on Imaging Systems and Techniques, 2008, pp. 216-220.

W.M. Laurent, J.M. Bonny, J.P. Renou, Imaging of water and fat fractions in high-field MRI with multiple slice chemical shift-selective inversion recovery. *Journal of Magnetic Resonance Imaging,* 2000, **12**, 488-496.

A. Lehnert, J. Machann, G. Helms, C.D. Claussen, F. Schick, Diffusion characteristics of large molecules assessed by proton MRS on a whole-body MR system. *Magnetic Resonance Imaging,* 2004, **22**, 39-46.

J. Ma, Dixon techniques for water and fat imaging, *J. Magn. Reson. Imaging,.* 2008, **3**, 543-58.

R. Mahdjoub, J. Molegnana, M.J. Seurin, A. Briguet, High resolution magnetic resonance imaging evaluation of cheese. *Journal of Food Science,* 2003, **68**, 1982-1984.

C. Papan, B. Boulat, S.S. Velan, S.E. Fraser, R.E. Jacobs, Two-dimensional and three-dimensional time-lapse microscopic magnetic resonance imaging of Xenopus gastrulation movements using intrinsic tissue-specific contrast. *Developmental Dynamics,* 2007, **236**, 494-501.

J. Qi, N.J. Olsen, R.R. Price, J.A. Winston, J.H. Park, Diffusion-weighted imaging of inflammatory myopathies: Polymyositis and dermatomyositis. *Journal of Magnetic Resonance Imaging,* 2008, **27**, 212-217.

J.M. Tingle, J.M. Pope, P.A. Baumgartner, V. Sarafis, Magnetic resonance imaging of fat and muscle distribution in meat. *International Journal of Food Science and Technology,* 1995, **30**, 437-446.

P. Walter, P. Cornillon, Lipid migration in two-phase chocolate systems investigated by NMR and DSC. *Food Research International,* 2002, **35**, 761-767.

MR MICROSCOPY OF FOOD FREEZING AND THAWING

Igor Serša, Franci Bajd, Ana Sepe

Jožef Stefan Institute, Jamova 39, 1000 Ljubljana, Slovenia

1 INTRODUCTION

Quality of conserved food critically depends on the food preservation method. While advantages of food preservation by freezing are apparent, the method has also disadvantages that are associated with the change of the food texture and consistence. Currently available technologies aim to freeze food instantly (immersion in very cold liquids) as this prevents formation of large ice crystals that may result in destroying the delicate cellular structure of the food [1-2]. Quick freezing is important, in particular in fruits and vegetables as they contain more water than other foods. More water makes food more susceptible to the formation of large ice crystals.

Microscopically, fast freezing induces intracellular ice formation, which plays an important role in the preservation of a cell structure, whereas slow freezing results mostly in an extracellular ice formation, which implies cell dehydration due to osmotic shock [3]. The freezing and thawing processes in fruits and vegetables were successfully traced by magnetic resonance imaging (MRI) in the previous studies [4-9]. The principal findings of the studies were that the MRI signal intensity changes during freezing/thawing were dependent on the portion of ice fraction, and that the relaxation and diffusion properties modified. Similar conclusions were drawn from the MRI studies on trout and bread dough samples [10-11]. However, only a few studies focused on continuous monitoring of food freezing and thawing by MRI were done [11].

In this study, magnetic resonance microscopy (MRM) was used to follow dynamics of slow food freezing and thawing process in vegetables and fruits, while MR relaxometry and diffusiometry were used to analyse food changes induced by slow and by fast freezing.

2 MATERIALS AND METHODS

2.1 Freezing dynamics

Food freezes because heat is transported from the food sample to the surroundings. This transport is driven by the temperature difference between the sample interior, which is during freezing at the freezing temperature T_f, and the external temperature T_e (the temperature of surroundings), which must be lower than the freezing temperature. The rate of the heat transfer is proportional to the temperature difference, to the coefficient of the thermal conductivity of the frozen food k and is inversely proportional to the thickness of the frozen food layer d. The major contribution to the transferred heat is the latent heat released during food freezing, while the minor contribution is due to the additional cooling of the frozen region associated with its thickening. Namely, as the frozen region thickens, the temperature gradient drops and so does the average sample temperature. The relation between the time t needed to freeze a layer of thickness d of a spherical sample with radius R is given by the following equation

$$t(d) = \frac{\rho L_f R^2}{6k(T_f - T_e)}\left(1 - \left(1 - \frac{d}{R}\right)^2\left(1 + \frac{2d}{R}\right)\right). \tag{1}$$

Here only the transfer of the latent heat was encountered. From Eq. 1 the freezing time may be obtained

$$t_f = \frac{\rho L_f R^2}{6k(T_f - T_e)}. \tag{2}$$

Obviously the freezing time is proportional to the sample radius squared R^2, to the sample density ρ, to the latent heat of freezing L_f and is inversely proportions to the thermal conductivity k and the temperature difference $(T_f - T_e)$.

2.2 NMR Freezing Apparatus and Sample Preparation

Food freezing and thawing was studied by a specially designed apparatus for freezing and thawing of food samples (Fig. 1). The apparatus consisted of a liquid nitrogen reservoir in which a 30 W heater was immersed. As the heater started heating, the boiling rate of nitrogen increased thus creating a flow of a cold nitrogen gas (the boiling temperature of nitrogen is -196 °C). The nitrogen gas was then directed into an NMR probe through a gas inlet tube with an additional heater and a temperature sensor. The sensor and the heater were connected to a temperature controller that was adjusting the heater power to heat the nitrogen gas to the selected temperature. This was in all freezing experiments set to -18 °C. The temperature response of the freezing apparatus was relatively slow as approximately 15 minutes was needed for the freezing apparatus to cool the probe with a sample for the room temperature of approximately 25 °C to -18 °C. This temperature (-18 °C) was maintained for additional 50 min, which was found enough to slow freeze any food sample of a diameter less than 2 cm. Food freezing was followed by thawing, which was started by turning off the heater immersed in liquid nitrogen and so stopping the flow of cold nitrogen gas into the NMR probe. In 90 minutes thawing time the probe and the sample temperature approached the room temperature within less than 1 °C difference.

In addition to the slow freezing experiments (as described above) fast freezing experiments were done too. In these, food samples were first immersed in liquid nitrogen for 10 minutes and then removed from it and allowed to warm to the room temperature (25°C) in additional 60 minutes.

Selected vegetables, fruits and meats were investigated. Specifically, freezing of carrots, raspberries and beef meats were studied. Carrot samples were prepared by cutting 2 cm long sections of carrot roots of approximately 2 cm in diameter, while raspberries were left intact and beef meat samples were 15 mm cubic shaped chunks of meat.

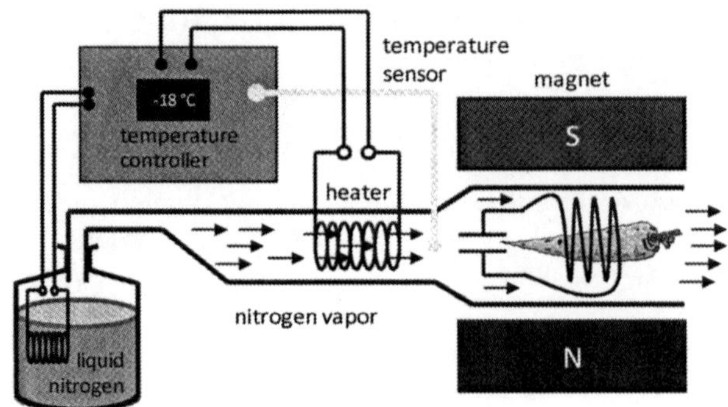

Figure 1   *A scheme of the NMR freezing apparatus.*

2.3 Magnetic Resonance and Optical Microscopy

MR study of food freezing and thawing was performed in an MR system consisting of a 2.35 T Oxford superconducting magnet, a Tecmag spectrometer and a thermoregulated 25 mm RF probe. In all experiments, slow as well as fast food freezing, the samples were before freezing and after thawing imaged by the conventional T_2-weighted spin-echo imaging sequence with echo time 28 ms and repetition time 1000 ms, by the multi spin-echo sequence with eight echoes, inter-echo time 18 ms and repetition time 4000 ms and by the diffusion-weighted imaging (DWI) sequence with b values 0, 99, 198, 298, 397, 495, 594 and 694 s/mm^2 (echo and repetition times were 28 ms and 1000 ms, respectively). Imaging field of view was 30 mm, image matrix and slice thickness 256 x 256 and 2 mm for the T_2-weighted sequence and 128 x 128 and 4 mm for the multi spin-echo and the DWI sequence. In all sequences images were acquired with two signal averages. Dynamical imaging of slow freezing and thawing was done by the same T_2-weighted spin-echo sequence as described above except the imaging matrix was reduced to 128 x 128 and no signal averaging was used so that the scan time was reduced to 132 s. In the slow freezing experiment the sample was all times in the magnet at exactly identical position. Dynamical imaging was started simultaneously with freezing. After 30 scans (66 minutes) freezing was stopped and in another 40 scans the sample was thawed and warmed to the room temperature.

Maps of the relaxation time T_2 and the apparent diffusion coefficient (ADC) were calculated from a series of eight T_2-weighetd images acquired by the multi spin-echo sequence and from a series of eight diffusion-weighted images acquired by the DWI sequence. The maps were calculated using the MRI Analysis Calculator, a plug-in of the ImageJ image processing software (NIH, Bethesda, Maryland, USA).

Some samples were examined fresh and after the freezing/thawing cycle under the optical microscope. The microscopy was performed by Nikon 80i Eclipse optical microscope equipped with a Nikon 20x Plan Fluor objective.

3 RESULTS AND DISCUSSION

An example of a slow freezing experiment is shown in Fig. 2b, which depicts a freezing dynamics of a carrot sample by a series of dynamically acquired T_2-weighetd images. Every fifth images of the first 60 images is shown so that images are 11 minutes apart. The temperature profile in Fig. 2a shows a time course of the RF probe (and the sample) temperature. After approximately 15 minutes the probe temperature reached equilibrium of -18 °C, which is associated with the start of sample freezing. This can be seen in the MR images as vanishing regions of the MR signal, i.e., frozen parts of a sample yield no signal detectable by MRI. As the freezing progressed more sample structures became signal voids. The freezing process progressed from outside to inside so that the last structures that froze were watery structures in the sample core (central cylinder of the carrot in the 4th image). The freezing process was the fastest at beginning when outer structures were freezing and then slowed with the progression toward the inner structures. Such dynamics is in a good agreement with the freezing model given by Eqs. 1 and 2. When the sample was completely frozen it yielded no MR signal (5th to 7th image). Sample thawing, that started at 66th minute (7th image), progressed from outside to inside. Thawed sample regions regained the MR signal and appear bright on MR images. Till 120th minute most of the sample was thawed, only a small triangular region extending from the centre remained partially frozen (11th image). It can be noted that the sample texture, that is clearly visible in a form of rays extending from the centre in MR images before freezing (1st and 2nd image) practically completely disappeared after thawing. This indicates a significant damage of the vegetable tissue during slow freezing, which is associated by the growth of large ice crystals that damaged cells. The cell damage could result also from the osmotic shock induced by the extracellular ice formation.

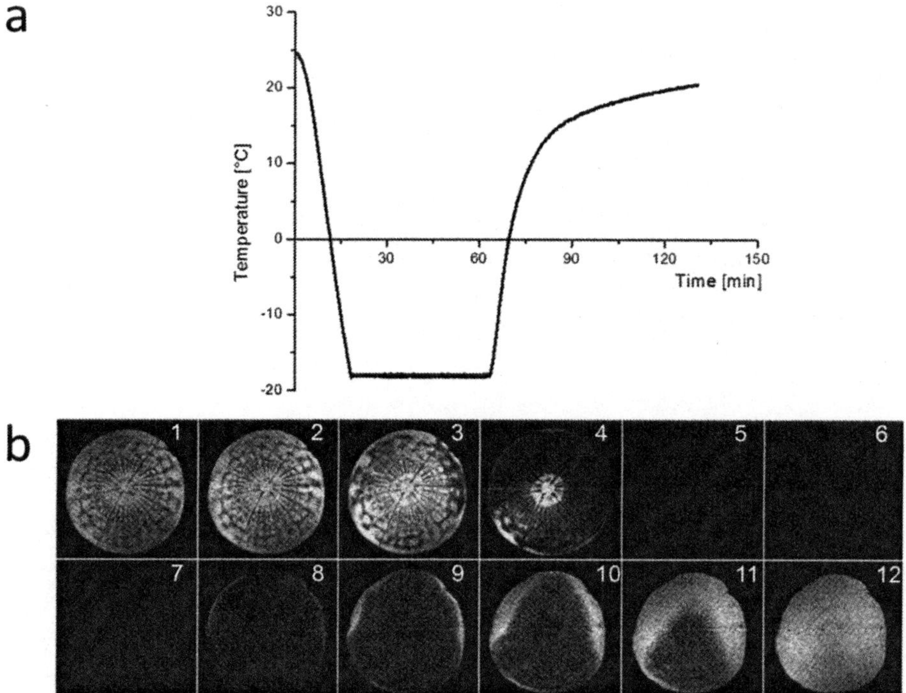

Figure 2 *Time profile of the RF probe temperature during slow freezing and thawing (a) and a corresponding series of dynamically acquired T2-weighted images (in 11 min intervals) of a carrot sample (b). Apparent is a complete loss of the sample structure after thawing.*

Perhaps, the freezing induced food changes can be seen even better in T_2 and ADC maps. Namely, both imaging modalities are quite sensitive to the appearance of the water in the tissue, in particular to water surroundings (T_2 map) and to water mobility (ADC map). The maps of the carrot sample along with its T_2-weigheted images are shown in Fig. 3.

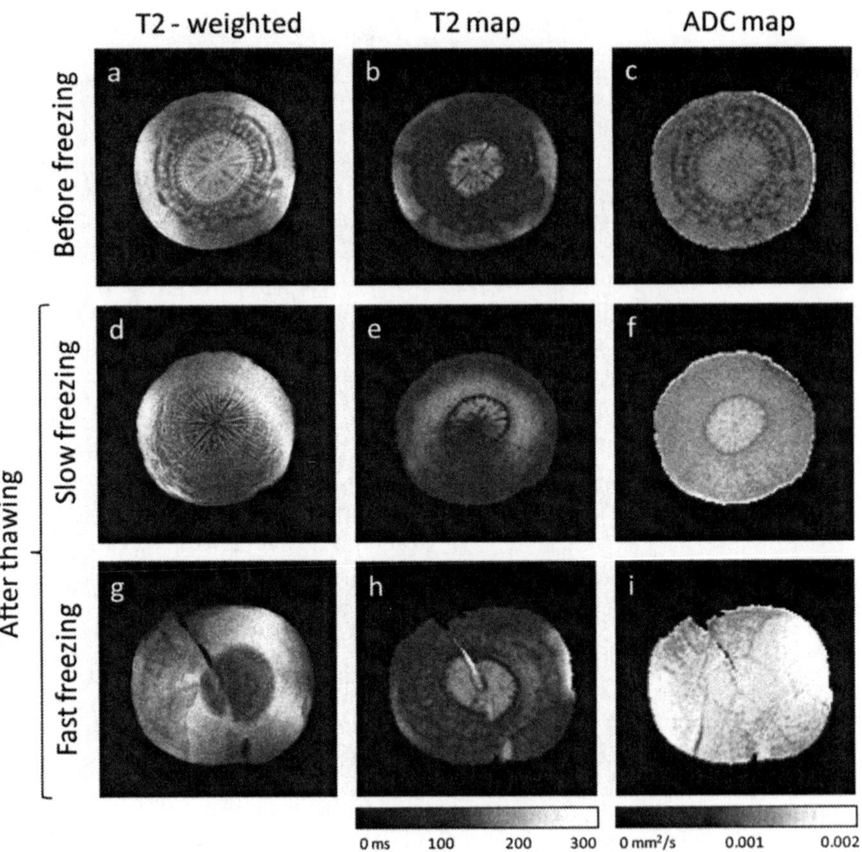

Figure 3   *T_2–weigheted images and T_2 and ADC maps of the carrot sample before freezing (slow and fast) and after thawing.*

All three imaging modalities (T_2-weigheted, T_2 map, ADC) show significant heterogeneity of the fresh sample (before freezing). The outer (cortex) region and the central cylinder (xylem) region of the carrot root have the highest T_2 and ADC values, while the intermediate region (secondary phloem with parenchyma) has almost twice shorter T_2 relaxation time and somewhat lower ADC. Freezing caused significant tissue damage that resulted in apparent texture loss (in T_2-weigheted images) and in an increase of T_2 and ADC values. The changes are the biggest in the intermediate region in which slow freezing resulted in almost doubled T_2 values, while fast freezing in the same region did not induce such big changes of T_2 and the sample texture is better preserved as can be seen in the corresponding T_2 map. Interesting is also a comparison between ADC maps that correspond to slow and to fast freezing. In both maps ADC values increased, which indicates that water mobility after the freezing/thawing cycle increased. However, the increase is bigger in fast freezing than in slow freezing (47% vs. 24% increase, data in Table 1) which is surprising and contra intuitive due the expected higher tissue damage in slow freezing in comparison to fast freezing. The result may be to some extent explained due to high water concentration in carrot and its heterogeneous structure that caused extreme mechanical tensions during fast freezing (because of these the sample cracked during fast freezing) that are most likely responsible for the excessive water drip of the thawed sample.

Table 1 *Average T_2 and ADC values of various food samples before and after (slow or fast) freezing.*

Average ADC values [10^{-9} m^2/s]

Food	Before freezing	After slow freezing	After fast freezing
Raspberry	1.22 ± 0.16	1.31 ± 0.16	1.39 ± 0.17
Carrot	1.07 ± 0.09	1.33 ± 0.10	1.57 ± 0.20
Beef meat	1.00 ± 0.11	1.04 ± 0.11	1.31 ± 0.15

Average $T2$ values [ms]

Food	Before freezing	After slow freezing	After fast freezing
Raspberry	178 ± 33	171 ± 26	153 ± 17
Carrot	125 ± 30	137 ±28	135 ± 23
Beef meat	65 ± 5	67 ± 6	63 ± 4

In addition to the carrot sample (as a representative of vegetables), the same analysis by T_2 and ADC mapping was applied also to raspberry (fruits) and beef meat (meats) samples. Results of the analyses are presented in Table 1 by average T_2 and ADC values that were obtained from corresponding T_2 and ADC maps, of the samples before and after slow or fast freezing. In all cases freezing increased ADC values (fast freezing more than slow freezing) and decreased the average T_2 value in the raspberry sample, increased it in the carrot sample and did not change it in the meat sample.

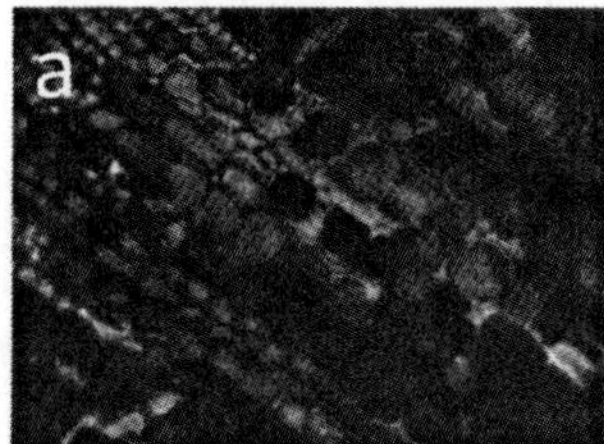
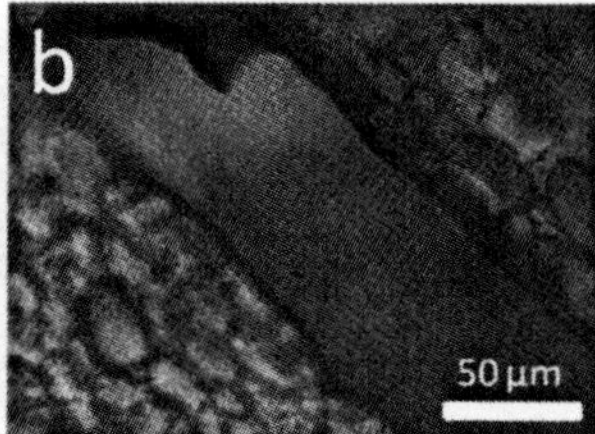

Figure 4 *Optical microscopy images of the carrot sample at 20-fold magnification. Images correspond to the fresh sample (a), to the slow frozen sample (b) and to the fast frozen sample (c).*

Figure 4 shows optical microscopy images of the carrot sample in the cortex region before and after freezing are shown. It can be seen that cells of the fresh sample are organized in a tight structure (a), while after the freezing/thawing cycle the structure is looser with large water filled spaces. These are considerably bigger in samples with slow freezing (b, more than 50 µm across) than in samples with fast freezing (c, arrow mark).

Freezing induced tissue damage could be significantly reduced by lowering the amount of water in the tissue or by an additional food treatment before freezing. For examples blanching is very helpful in reduction of the water drip and other freezing induced changes in carrots.

4 CONCLUSION

This study demonstrated that MR microscopy is a powerful tool in analysis of freezing induced changes in food. Food freezing experiments can be performed by a relatively simple NMR freezing apparatus that enables dynamical MRI studies of freezing and thawing of small food samples. As frozen food gives no detectable MRI signal, MRI can be used for accurate determination of proportions between fresh and frozen food. Freezing induced tissue damage can be efficiently detected by T_2 and ADC mapping, which are sensitive to changes of the environment and mobility of water.

Acknowledgment

The authors thank to Davorin Kotnik for technical assistance. The study was financially supported by the J4-2053 Slovenian Research Agency grant and the EN-FIST Centre of Excellence.

References

[1] Holdsworth SD. *The effect of rate of freezing on the quality of frozen foods.* [S.l.]: F.V.P.R.A.; 1968. pp. 52. p.
[2] Van Arsdel WB. *Quality and stability of frozen foods : time-temperature tolerance and its significance.* Wiley-Interscience; 1969. 384p.,ill.,24cm. p.
[3] Chua KJ, Chou SK. On the study of the freeze-thaw thermal process of a biological system. *Applied Thermal Engineering,* 2009;**29**(17-18):3696-3709.
[4] Hills BP, Goncalves O, Harrison M, Godward J. Real time investigation of the freezing of raw potato by NMR microimaging. *Magnetic resonance in chemistry,* 1997;**35**(13):S29-S36.
[5] Kerr WL, Clark CJ, McCarthy MJ, de Ropp JS. Freezing effects in fruit tissue of kiwifruit observed by magnetic resonance imaging *Scientia Horticulturae,* 1997;**69**(3-4):169-179.
[6] Kerr WL, Kauten RJ, McCarthy MJ, Reid DS. Monitoring the Formation of Ice During Food Freezing by Magnetic Resonance Imaging LWT - *Food Science and Technology,* 1998;**31**(3):215-220.
[7] Koizumi M, Naito S, Haishi T, Utsuzawa S, Ishida N, Kano H. Thawing of frozen vegetables observed by a small dedicated MRI for food research. *Magnetic Resonance Imaging* 2006;**24**(8):1111-1119.

[8] Hills BP, Nott KP. NMR studies of water compartmentation in carrot parenchyma tissue during drying and freezing *Applied magnetic resonance,* 1999;**17**(4):521-535.

[9] Gamble GR. Non-Invasive Determination of Freezing Effects in Blueberry Fruit Tissue by Magnetic Resonance Imaging. *Journal of Food Science,* 1994;**59**(3):571-573.

[10] Nott KP, Evans SD, Hall LD. Quantitative magnetic resonance imaging of fresh and frozen-thawed trout. *Magnetic Resonance Imaging,* 1999;**17**(3):445-455.

[11] Lucas T, Grenier A, Quellec S, Le Bail A, Davenel A. MRI quantification of ice gradients in dough during freezing or thawing processes. *Journal of Food Engineering,* 2005;**71**(1):98-108.